| DATE DUE | | | |
|----------|--|--|--|
| Nov13 '80 | | | |
| | | | |
| | | | |
| | | | |
| | | | |
| | | | |
| | | | |
| | | | |
| | | | |
| | | | |
| | | | |
| | | | |

MINERAL CYCLING
IN A TROPICAL MOIST FOREST ECOSYSTEM

# Mineral Cycling
## in a
# Tropical Moist Forest
## Ecosystem

FRANK B. GOLLEY, JOHN T. McGINNIS,
RICHARD G. CLEMENTS, GEORGE I. CHILD, AND
MICHAEL J. DUEVER

WITH CONTRIBUTIONS FROM JAMES DUKE,
JOHN EWEL, CLAYTON GIST

*University of Georgia Press*
*Athens*

Library of Congress Catalog Card Number: 73-89717
International Standard Book Number: 0-8203-0344-5

The University of Georgia Press, Athens 30602

Printed in the United States of America

# Dedicatoria

Al sentarnos a la vera del camino para aquilatar nuestros logros e informar los resultados obtenidos, dedicamos esta publicación científica a los pueblos hermanos de Panamá y Colombia. Su hospitalidad hizo de nuestra visita a sus paises una de eterna recordación; su nobleza y bondad suavizó el fragor de la lucha investigativa; su cooperación aumentó la productividad de nuestra investigación. Deseamos y confiamos que estos estudios contribuyan a la utilización productiva y al manejo efectivo de sus magnificas tierras y bosques.

# Dedication

This research publication is dedicated to the peoples of the Republics of Panama and Colombia: their hospitality made our visits to their countries a pleasant experience; their kindness made our work easier; their assistance made our studies productive. It is our wish that these investigations will contribute to the wise utilization and management of their magnificent lands and forests.

# Table of Contents

# Figures

# Tables

# Foreword

The tropical rain forest is perhaps the prime example of a self-maintaining, homeostatic ecosystem. For an incredibly long period of time it has existed in balance with its environment and it would continue to do so indefinitely but for the impact of civilized man which is upsetting this balance so completely that the continued existence of the system for more than a few decades is very doubtful.

The aspect of the rain forest in which its balanced homeostatic character is most clearly shown is its mineral cycle, which is organized in such a way that luxuriant vegetation can exist on soils so poor in plant nutrients that once the forest is destroyed permanent agriculture is difficult or impossible to maintain. Forest maintenance under these circumstances is possible because the system's mineral capital is recycled so efficiently that in spite of year-round heavy rainfall only very small quantities of mineral ions are carried away in the drainage water.

The first realization that the clue to the paradox of a luxuriant forest and an infertile soil lies in the almost closed mineral cycle came some forty years ago from the studies of H. Walter and G. Milne in Africa and of F. Hardy in the West Indies, summarized in my own book *The Tropical Rain Forest* (1952, reprinted 1964, pp. 219-220). Further important contributions on the mineral cycle have appeared in more recent years, but to obtain a complete quantitative picture of the cycle one must still overcome formidable practical and techinical difficulties.

It is fortunate that thanks to the stimulus and opportunities provided by the Battelle Institute's Sea-level Canal Project it has been possible to study the problem more thoroughly than ever before. There is still a wide gap between what is theoretically desirable and what is practically possible, and not all the difficulties have been overcome, but this book includes the most complete set of data on the mineral cycle in tropical forests that has yet been collected.

For this, Frank Golley and his team must be warmly congratulated. Their work is a notable contribution to understanding the tropical forest as an ecosystem and it is likely to be of considerable practical value, not only as part of the feasibility studies for the canal project, but as a valuable basis for the efforts to devise more efficient systems for raising the food crops so urgently demanded by present economic and social conditions in the humid tropics.

PAUL RICHARDS

# Preface

In January, 1966, a research team from the University of Georgia was invited to Columbus, Ohio, by Battelle Memorial Institute to contribute to the planning of bioenvironmental and radiological safety studies of the feasibility of the excavation of a sea-level canal and, further, the feasibility of using explosives for the excavation process. This meeting began a highly productive and enjoyable relationship with the Atomic Energy Commission, Battelle Columbus, and its subcontractors. The studies led us to a new literature, many new friends, for some of us a new language, and most importantly a new way of thinking about the land. Many of our colleagues expressed grave doubts that we could gain sufficient insight into the dynamics of tropical forests to predict the consequences of excavation with nuclear explosives in the one year available for field study. Our response was that ecology was being asked to contribute to an exceedingly important and exciting engineering program: a program that, if successful, would permit man to change the shape of the landscape. If ecology was to be a practical as well as a theoretical science, it had to compromise with the ideals of scientific investigation and provide useful, if preliminary and possibly crude, answers for engineers and planners.

A number of specialists participated in the feasibility study, and the results of their combined efforts are most appropriately reported in various publications of Battelle Memorial Institute. The environmental data which are directly applicable to feasibility of canal construction with nuclear explosives will be presented in their reports. However, our data are also of direct ecological interest, with applications to theories of tropical forest development and ecosystem mineral cycling. This volume is designed to present these findings from a purely ecological viewpoint. Information on tropical forests and on mineral cycling has great relevance to human welfare, and for this reason we have directed our publication toward the applied scientist as well as the ecologist.

The study proceeded in three phases. Phase I, from June, 1966, to December, 1966, consisted of library research and a reconnaissance trip to Panama, under the direction of John T. McGinnis. Phase II field work began in January, 1967, and was coordinated by Richard Clements, with supervision of field crews by George Child and Michael Duever. Andrea Duever operated the laboratory in the Canal Zone for initial reduction of samples. Biological samples were prepared in this laboratory and shipped to the Savannah River Ecology Laboratory for analysis by Helen Morrissett and Susan Wagner, under the direction of Robert Beyers. Phase III, from December, 1967 to June, 1968, consisted of analysis of data and preparation of publications and reports including this volume. These latter phases were under the direction of Frank B. Golley.

Thus this volume is the result of an interdisciplinary team effort. The basic, guiding concept of the approach is the ecosystem, in the sense of Tansley and E. P. Odum. The research on which the volume is based proceeded systematically from this concept as we considered ecosystem components or subunits which we assumed to be important in forest mineral cycling. As we worked with the forest we tested our assumptions and consequently the list of significant components changed; nevertheless, the study progressed directly toward the objectives because of the overall concept of the ecosystem and the central control through the research team. Is there any other way to study complex systems?

It will be obvious to the reader that the study is not definitive. Many data are still lacking and many questions remain unanswered. Further, the long period intervening between the first draft of this monograph (1968) and its publication has meant that many reports on mineral cycling in tropical systems which were unavailable when the study was completed had to be incorporated within an internally consistent manuscript. Even so, we hope that this description of mineral cycling in a tropical forest will be of use in the development of sound land management practices in the tropical forest.

FRANK B. GOLLEY

# Acknowledgments

The University of Georgia research program was completed with the assistance of many persons. Our problems of logistics, analysis and preparation of the reports often were solved only by the attention of individuals well beyond their specified duties and responsibilities. Those who assisted us in special ways during some phase of the project, listed alphabetically by institutions were: Battelle Memorial Institute, Columbus, Ohio: Richard S. Davidson, William E. Martin, Roger L. Evans; Battelle Memorial Institute, Balboa, Canal Zone: F. Webster McBryde, James A. Duke, Mrs. Elinor K. Willis, Mrs. T. Victoria McGrath, Mrs. Barbara Child, Mrs. Diane A. Kutny; Battelle Memorial Institute—Pacific Northwest Laboratories: William L. Templeton; Cia de Navegacion Latina S. A.: Joseph Cecil; Corps of Engineers: Col. Alex G. Sutton, Jr., Col. Harold G. Stacy, Maj. Richard R. Manahan, Maj. Van R. Bonnewitz; Department of the Army: Cecil E. Dodson, Jr., Charles G. Canning; Government of the Republic of Panama: Simon Quirós Guardia; Pan Canal Company: A. O. Icaza; Pan Canal Company, Hydrology Section: Theodore C. Henter, Frank H. Robinson, John R. Robinson; Missouri Botanical Garden: John D. Dwyer; Puerto Rico Nuclear Center: Frank G. Lowman; Tropic Test Center (DOD); Robert S. Hutton, George W. Gauger; U.S. Army (Medical Corps); Maj. Bruce F. Eldridge, Cpt. Reid R. Gerhardt, Cpt. David G. Young, Cpt. Karl K. Longley; U.S. Air Force: Lt. Col. C. O. Little, Jr.; U.S. Atomic Energy Commission NVOO: A. W. Klement, Jr., Jared Davis; USAEC, OICS: J. E. Hebert, Lee L. Fisher, Virgil W. Luckett; U.S. Public Health Service: Roy Evans; University of Florida: John F. Gamble, John Ewel, Raymond B. Reneau, Jr., M. William Silvey, Miss Mary Perrin; University of Georgia: Mrs. Valeta Morris, Mrs. Susan Wagner, Mrs. Elaine Mahoney, Mrs. Ann Parrot, Bernard Patten, Wade Nutter, Robert Beyers, J.B. Jones, Mrs. Helen Morrissett, Andrea Duever, Mrs. Thelma Richardson, Mrs. Priscilla Golley, Mrs. Jimmie Wagner, Mrs. Cheryl Coker; University of Panama: Dra. Reina Torres de Arauz, Sr. Amado Arauz. Finally, Peter McGrath who took many of the photographs.

In addition, numerous employees of the Panama Canal Company and residents of Darien and San Blas Provinces made our work easier in many ways. We express our appreciation to all of these persons collectively.

We are also grateful to Philip Johnson, Dale Cole, D. A. Crossley, David E. Reichle, and Helmut Lieth for reading all or part of the manuscript.

The project was supported by a contract between Battelle Memorial Institute, Columbus, and the U.S. Atomic Energy Commission, At (26-1)-171.

# Resumen

El presente informe describe los ciclos de los elementos químicos esenciales en el bosque húmedo tropical de bajura en la Provincia de Darién, República de Panamá. El trabajo fue una parte del Estudio sobre el Canal Interoceánio en 1966-1967.

El objetivo básico del estudio fue desarollar un modelo de ecosistema del bosque húmedo tropical de bajura que describiera los movimientos de los elementos dentro de los componentes bióticos del bosque y entre la biota y la atmósfera, el suelo y el agua. La descripción del ciclo en el bosque podría ser comparada con datos semejantes de otros bosques en la Provincia de Darién y con bosques de otras partes del mundo. Se eligieron dos sitios de estudio cerca de Santa Fe, Provincia de Darién. En cada uno se tomó un lote de muestreo de un quarto de hectárea. Una muestra fue estudiada en el tiempo seco y la otra durante el periodo de lluvias. Datos sobre la precipitación y sobre hojarasca fueron obtenidos cerca de Santa Fe durante un año. La mayor parte de la vegetación fue cosechada en cada lote estimándose la biomasa total a partir de los datos de cosecha. Los componentes del bosque muestreados en este proyecto fueron las hojas, los tallos, las raices, la hojarasca, los frutos y flores, el suelo hasta 30 cm de profundidad, los herbívoros, los carnívoros y los detritívoros. El contenido químico de la biomasa, de las precipitaciones, del agua del rio, del suelo y de la hojarasca fue determinado con métodos químicos apropriados para el fósforo, potasio, calcio, magnesio, sodio, aluminio, cobalto, bario, cobre, cesio, hierro, manganeso, molibdeno, níquel, estroncio, titanio, plomo, zinc. También fue determinada la cantidad de nitrógeno.

Estos datos permitieron calcular el inventario de los elementos en los componentes del bosque (Cuadro 2.19) y el flujo de los elementos entre los componentes (Cuadro 3.21) usando un modelo determinístico linear (Figura 1.3). El "uptake" annual en el bosque húmedo tropical de bajura es cercano al 10 por ciento o menos del inventario mineral en los componentes orgánicos del bosque; y el "turnover" (que es el inventario mineral dividido por la hojarasca annual) es siempre menos de 150 años y en general menos de 20 años. Estos datos indican la repetición del ciclo de ciertos elementos esenciales en este tipo de bosque.

Para la mayor parte de los elementos, el "input" en el bosque debido
a las precipitaciones casi iguala el "output" de la descarga en el rio.
Solamente las pérdidas de calcio, magnesio y sodió son significativa-
mente grandes. Estos elementos son recobrados mediante procesos de
mineralización en el suelo. Si este bosque fuera cortado, necesitaría el
reemplazo del inventario depositado en la vegetación y en los animales,
desde el suelo o desde los fertilizantes. Potasio y fósforo parecen ser
los elementos más limitantes para el crecimiento del bosque. Ambos se
encuentran en grandes cantidades en el inventario orgánico, ambos
ciclan rápidamente en el ecosistema, y ambos se encuentran en pe-
queñas cantidades en el suelo. Estos elementos se encuentran en más
del 80 por ciento en la parte biótica del sistema, en contraste con el
calcio que concurre en más del 80 por ciento en el suelo. Al respecto,
el bosque húmedo tropical de bajura es similar al bosque muy húmedo
premontano, al bosque inundable y al manglar, en Panama.

Las decisiones de manejo respecto a la utilización de las tierras de
los bosques tropicales, para la producción agrícola, la ingenieria
forestal, el pastoreo y otros usos, deberán basarse en los datos de
los recursos nutritivos y en las cantidades disponibles—como algunos
descritos en este informe—, y también en factores sociales, econó-
micos y otros. En ciertos casos los recursos nutritivos podrán limitar el
manejo de la tierra.

# I
# INTRODUCTION

*What attracted us chiefly were the colossal trees. The general run of trees had not remarkably thick stems; the great and uniform height to which they grow without emitting a branch, was a much more noticeable feature than their thickness. . . .*

The tropical world lies approximately between the latitudes 23½° N and 23½° S. It contains examples of many vegetation types found on the planet, but it is unique in having large areas of luxuriant forests. These forests are popularly called tropical rain forests and are characterized by a large variety of species, great height of the trees, and many vines, epiphytes, palms and other plants unusual in temperate forests. On one acre as many as a hundred species of trees may grow, with individual trees of any one species often widely scattered throughout the forest. Animals are equally profuse, and since many live in the canopy of the trees, high above the soil, they are often difficult to observe and study.

Where rainfall is heavy throughout the year, rain forest is the predominant vegetation. Although a wet season and a dry season are almost universally recognized by the forest inhabitants, in some areas the dry season is actually only a season of slightly less rain. In other areas where the dry season is longer in duration and the soil has the opportunity to dry out, deciduous species which drop their leaves during the drier period occur. In this case, the rain forest changes to a semideciduous or seasonal rain forest; it still possesses the variety of species and other characteristics of rain forests, but also shows a strong seasonal response by plants and animals.

While these forests are found in the Americas, Africa, Asia, and Australia, the largest single area of tropical forest in the world is found in and around the Amazon Basin. This vast expanse of trees extends south to about latitude 25° S on the east coast of South America, north to the Caribbean and west to the Andes. Rain forests and semideciduous rain forests also extend north to Mexico through the lowlands of Central America.

About 10 percent of the world's land area, occupied by about 28 percent of the world's population, is potentially covered by these forests. Certain parts of this area, such as the Amazon Basin and Central America, have relatively few inhabitants. In these regions many people earn their livelihood through shifting agriculture, a form of agriculture in which forest is cut, burned, and crops planted in the ash among the charred stumps and branches. Fields prepared in this way are useful for only a few years because of declining fertility, competition of crop plants with weeds, and consumption of crops by insects. Then another section of forest must be cut and reduced to ash. In other rain forest regions, the hot, well-watered alluvial plains have been cleared of trees and converted to fields of rice and other crops. It is in these areas in southeast Asia and India that dense populations of man occur. Discounting these populations, only about 5 percent of the world's population occurs in rain forest areas. Especially in the Americas, the forests form a vast reserve potentially available for settlement, agriculture and industry. Since the rain forest in the Americas is bordered by highlands and urban areas groaning under the load of excess population, they have attracted the attention of the developer.

Successful development of the land rests upon a sound knowledge of the available human and natural resources, and the factors limiting production. Among the information needed for land planning and management are data on the productivity and nutrient availability of the fields and forests. Agricultural and forest production depends mainly upon adequate supplies of sunlight, water, and nutrients. Nutrients are the chemical materials from which living matter is built and maintained, and are derived from the soil, water, and atmosphere. Green plants take up some of these essential chemicals from the soil through the roots. These materials are then translocated through the plant and incorporated into its tissues. When the plant or plant part dies, decomposition of the matter begins and the elements contained in it are returned to the soil where they are again available. An alternate pathway occurs when plant material is eaten by an animal; in this case, the return of the chemical elements to the soil is delayed until they are excreted or the animal dies. In either case, the essential nutrients may cycle over and over through the biological system. While the nutrients cycle in a variety of chemical forms, it is convenient to follow the pathways of the separate elements of the periodic table as if they moved separately through the system. This abstract analysis of chemistry of ecosystems is called "mineral cycling" in the ecological literature.

This description of mineral cycling tells only a part of the story. Cycling is never completely efficient; thus there is leakage from the

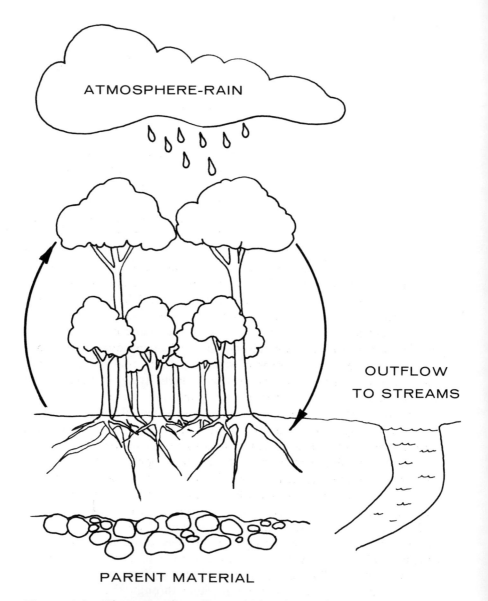

ATMOSPHERE-RAIN

OUTFLOW
TO STREAMS

PARENT MATERIAL

Figure 1.1 Diagram of cycling within the active ecological community. The community consists of vegetation, animals, and soil. Inputs to this community are from the atmosphere and rain and to the subsoil from the parent material. Output is to the stream.

biological-soil system which must be made up from the soil parent material, the atmosphere, or rain wash (fig. 1.1). There is a continuous source of nutrients to the community through chemicals in the rain water and from chemical weathering of soil minerals, just as there is a continuous loss of nutrients in water moving through the plants and soil to streams and rivers. We would expect that the chemical kinetics of communities would vary with rainfall and other environmental factors, soil parent material and with the nature of the vegetation. Some types of vegetation, such as agricultural fields with a changing vegetative cover, might hold nutrients less effectively than forests. Tropical forests occurring in high rainfall environments might be especially susceptible to a nutrient deficit since the soil is often weathered to a great depth and may have less recharge capacity, and the available nutrients could be leached easily from the soil in the heavy rainfall.

Actually there is relatively little information on nutrient cycling in tropical forests. The information on the chemical content of tropical vegetation has been summarized by Rodin and Bazilevich (1967); studies of the dynamics of the mineral cycling include those of Greenland and Kowal (1960), Laudelout and Meyer (1954), Nye (1961), Rozanov and Rozanova (1964), Tsutsumi et al. (1967), Odum and Pigeon (1970), and Stark (1971). While the information is quite limited, certain patterns are nevertheless suggested by the studies. First, the uptake and return of nutrients may be greater per year in tropical forests than in other types of vegetation (Rodin and Bazilevich, 1967; Dommerques, 1963). Second, a larger proportion of the entire chemical inventory of the system is held in the vegetation (Rodin and Bazilevich, 1967). Third, in tropical forests the percentage of the vegetation in green parts, the proportion lost per year as litter, and the rate of decomposition of the litter are all greater than in temperate forests (Rodin and Bazilevich, 1967; Dommerques, 1963). Fourth, the rate of uptake is strongly influenced by the rate of evapotranspiration (Odum, 1970).

These general observations suggest that tropical forests have a large mass of chemical elements held in the wood of the trunks, branches, and roots which move relatively slowly through the system. Also, a smaller mass associated with the green parts and the litter cycles rapidly between the soil and the plants. There is also a suggestion that the reserve of elements in the soil is relatively small and there is relatively little leakage from the system—a situation that Pomeroy (1970) also reports for coral reefs. We will test these generalizations with the observations made in this study.

There are more than one hundred chemical elements that might occur in the environment. The elements of interest in mineral cycling are those which are especially reactive given the normal conditions of temperature and pressure on the earth's surface. The basic six are carbon, hydrogen, oxygen, nitrogen, sulfur, and phosphorus. Deevey (1970) points out that no elements lighter than iron and cobalt (numbers 26 and 27 in the periodic table) are unimportant in the biosphere and that beyond copper (number 29) there are few conspicuously reactive elements. Hydrogen, oxygen, carbon, nitrogen, and sulfur are water soluble and volatile so that they move to the biosphere by way of the atmosphere or hydrosphere. Other elements such as phosphorus, sodium, calcium, magnesium, and iron move mainly through water.

The most abundant elements in the biosphere are oxygen, carbon, and hydrogen which occur in amounts greater than $10^4$ kilograms per hectare (Deevey, 1970). These plus nitrogen, calcium, potassium, magnesium, sulfur, phosphorus, iron, manganese, aluminum, boron, chlorine, copper, molybdenum, and zinc are essential for plant growth

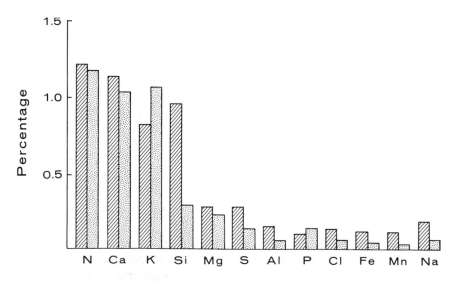

Figure 1.2. Comparison of the elemental composition of tropical and temperate forests. The data are from Rodin and Bazilevich (1967) and represent twenty or more separate samples of forests or species. The percentage composition based on dry weight is shown for tropical forest by open bars and for temperate forest by closed bars.

(Gilbert, 1957; Underwood, 1965). Cobalt and sodium are essential for animals. Nitrogen, calcium, potassium, silicon, magnesium, and sulfur occur in quantities of more than about 100 kilograms per hectare in the biosphere (Deevey, 1970). The remaining elements are present in trace amounts.

The distribution of abundances is not constant for all types of communities. For example, the percentage composition of the more abundant elements, excluding carbon, hydrogen, and oxygen, differs for tropical and temperate forest (fig. 1.2). Tropical forests contain larger percentages of silicon, magnesium, sulfur, and trace elements, but smaller quantities of potassium, than temperate forests. The reasons for differences such as these also will be explored later.

In the present research program, it was not possible to examine the entire suite of elements. Not only were our analytical capabilities insufficient to detect all the chemicals of interest, but the funding agencies for the interoceanic canal study which supported the work were interested primarily in those elements which might become radioactive and be a problem in the canal construction and operation. For these reasons the elements studied here—nitrogen, phosphorus, potassium, calcium, sodium, aluminum, barium, boron, cobalt, cesium, copper, iron, manganese, molybdenum, nickel, lead, strontium, titanium, and zinc—are a strange mixture. They include essential elements and nonessential elements that might occur as trace contaminants in biological tissues. Nevertheless, they provide an unusually broad spectrum for evaluation of the nutrient condition of the forests.

THE STUDY

The basic concept that underlies this study of tropical forest mineral cycling is that of the *ecosystem*. As used here, ecosystem means the complex of plants and animals living together in an ecological community and interacting with the physical and chemical environment. We assume that the ecosystem can be an object of study and can be examined analytically as can an individual, a cell, or a population. Clearly ecosystem science is in its infancy; even so, we submit that a most practical and efficient way to understand and manage landscapes is through the application of ecosystem principles. This study explores ecosystem principles concerned with nutrient cycling.

The study of ecosystems usually proceeds in a standard way. First, the system is identified and arbitrarily defined. Usually this definition considers its shape, mass, and horizontal distribution, as well as its specific functions in the biosphere. Next, the community is divided

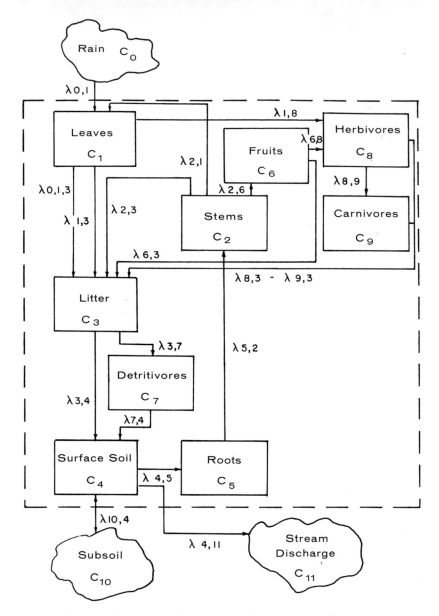

Figure 1.3. A diagrammatic model of a tropical forest ecosystem. The diagram depicts the nine organic compartments and illustrates the flow of minerals through the system. The compartments are labeled as $C_1$, $C_2$ . . . $C_{11}$ . $C_0$ is the atmosphere compartment, $C_{10}$ the subsoil and $C_{11}$ the river. Transfers between compartments are identified by $\lambda$, with subscripts indicating the compartment the flow was from and the compartment the flow goes to. Forest compartments are represented by a rectangle, compartments external to the system are represented by an amorphous figure.

into a limited number of groupings which are significant in terms of the structural and functional definition of the system. At this first step these groupings are seldom single taxa, such as genera or species. More often they represent groups of animals or plants which carry out some function in a common way. For example, the tropical forests can be divided into green plants, soil, litter, plant-eating animals, and so on (fig. 1.3). The character of each of these components is studied in the field where it may be necessary to determine the taxa of plants or animals in the component, their distribution over space and time, and their numbers, weight, size, and shape. Finally, the relationships among the components are determined. These couplings may include the transfer of chemicals or energy or chemical and behavioral interactions between components. In this study we were interested in the chemical kinetics of the tropical forest, therefore we determined the quantities of elements in each forest component and the rate of exchange between components. Each forest was divided into nine components (fig. 1.3). Exchange rates were measured only for the most abundant forest type in the Darien Province of Panama, the Tropical Moist forest, because of the extreme difficulty of making continuous observations in a remote area. Standing crop information was collected in Premontane Wet, Mangrove, Riverine, and successional forests in the Darien and in adjacent areas of northwestern Colombia in addition to Tropical Moist forest. The field work was carried out in 1967.

Our standard procedure in each of the forest types was to select a site which seemed typical of the general type and was reasonably close to a support base. At this site a one-quarter-hectare-square plot was located and marked with tape. The underbrush was removed next and was weighed. Then all of the trees were counted, mapped, and their diameter at breast height (DBH) determined (fig. 1.4). With this information a diameter profile of the plot could be drawn (fig. 1.5), which represented the distribution of the mass of the forest. A 10 percent sample of each diameter class was selected at random and the trees were cut and weighed. Leaves, branches, trunks, and roots were piled separately and random samples were taken from each for chemical analysis. These samples were bagged in plastic, shipped to the Canal Zone laboratory where they were oven-dried and ground to a powder in a Wiley mill. Samples of the powder were shipped to the University of Georgia for extraction with sulfuric-nitric-perchloric acid and were read on an atomic absorption spectrophotometer or were ashed and read on an emission spectrophotometer. Soils were

Figure 1.4.    Measurement of tree diameter for construction of the
diameter profile for the study plot. (Photograph by the authors.)

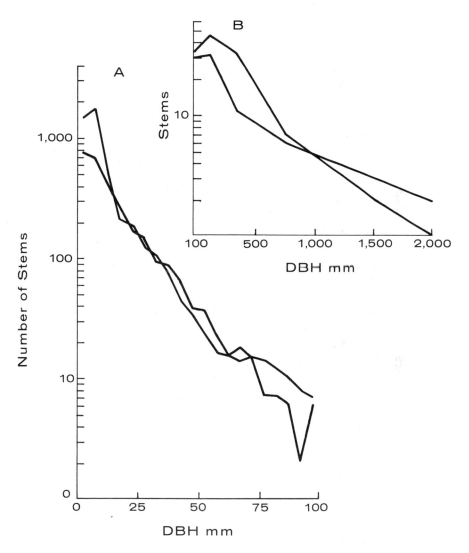

Figure 1.5. Diameter-abundance curves for two stands of Tropical Moist forest in Darien, Panama. The relationships illustrate the distribution of large and small stems in the forest and were the basis for the sampling of the organic mass. Stems from 1 to 99 mm DBH appear in graph A; and those 100 to 2,000 mm in graph B.

sampled at regular locations on the quarter-hectare plot to a depth of 30 cm. Animals also were sampled at the time of harvest of the vegetation. These data provided a knowledge of the standing crop of elements.

The transfer of elements between compartments was determined mainly for Tropical Moist forest in three steps. In this forest we measured litter fall throughout one year, rainfall and the chemical content of rain water, and the character of the river water draining the forest area. The combination of these and the standing crop data in a systems model permitted the calculation of other transfers and losses and completed the description of the cycling in the forest. It was unfortunate that all components and transfers could not be studied with equal intensity. We have tried to discuss assumptions and/or weaknesses in the data in the appropriate place in the text. We do not feel that these limitations introduce grave error into the overall conclusions, but they do limit the predictive capability of the models, and also pinpoint problems that require attention in future studies.

According to our procedure, the report begins with a conventional description of the physiognomy of Tropical Moist forest ecosystem. Following the biological description is an ecological description of the standing crops of organic matter and nutrients in plants and animals. The next chapters describe litter fall, rain wash from the canopy, the water budget, and end with a model of nutrient cycling in the Tropical Moist forest. The final chapters compare the structure and function of the Tropical Moist forest with other tropical and temperate forest systems. Detailed data and species lists are in appendices.

# II

# THE STRUCTURE
# OF THE TROPICAL
# MOIST FOREST

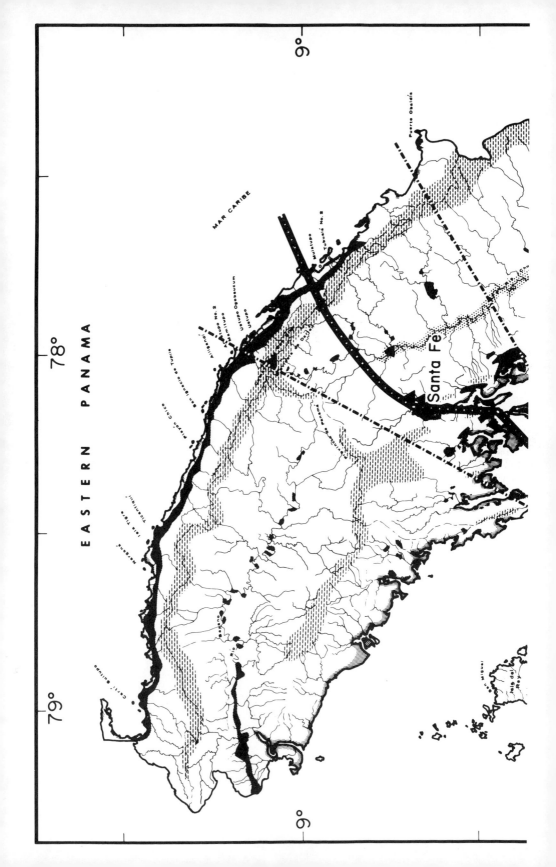

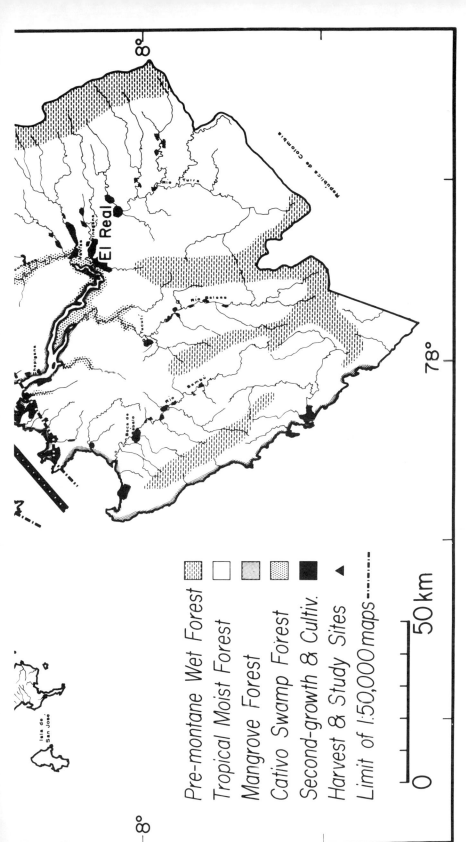

Figure 2.1. The distribution of vegetation types in eastern Panama. The area of study is indicated by dashed lines.

*There is a grandeur and solemnity in the tropical forest, but little of beauty or brilliancy of color. The huge buttress trees, the fissured trunks, the extraordinary air roots, the twisted and wrinkled climbers, and the elegant palms, are what strike the attention and fill the mind with admiration and surprise and awe.*

ALFRED RUSSEL WALLACE

CHAPTER 1

# Description of the Forest

The Darien Province of the Republic of Panama is characterized as a broad, flat valley bisected by the Chucunaque and tributary rivers, with the Caribbean coastal mountains on the east and a low Pacific coastal range on the west (fig. 2.1). On a flight over the province on TASA or ADSA, the local airlines, the viewer is impressed by the flat expanse of forest stretching mile after mile (fig. 2.2) broken only by the rivers. The historically minded traveller recalls that this is the area from which Balboa discovered the Pacific Ocean. The biologist or geologist, thinking in a longer time span, recalls the land-sea barrier between North and South America which prevented or permitted the mixing of the land faunas of the continents throughout geological history. The forests, as seen from the air, are typified by two emergent trees, the cuipo *(Cavanillesia platanifolia)* and the bongo *(Ceiba pentandra),* both of which lose their leaves during the dry period of January to March. The naked branches of the emergent trees are one of the most striking features of the forest of this area of Panama (fig. 2.3).

The forests of the Darien have been surveyed by Pittier (1918), Lamb (1953), and Holdridge and Budowski (1956). The latter two authors prepared a map of the plant formations of Panama, classifying the area of study in the Chucunaque River valley as transitional Tropical Dry forest or Tropical Moist forest. They stated, "Ordinarily transition areas between two life zones are narrow and are not mapped separately, but in Darien Province we found such extensive zones of transition that we mapped these out separately. In such transition areas, there is a mixture of the tree species from each formation plus certain species which are typically transition species. The interesting tree, cuipo . . . , is one of the latter. When one crosses the isthmus of Panama near the Zone, this huge tree is found fairly well restricted to the edges of the moist and dry forest in a narrow strip. However, in the Darien, this species occurs over a very wide section indicating clearly that this whole area is a transition belt." Later, however, Budowski (1966) classes this forest as successional. He states, "equally

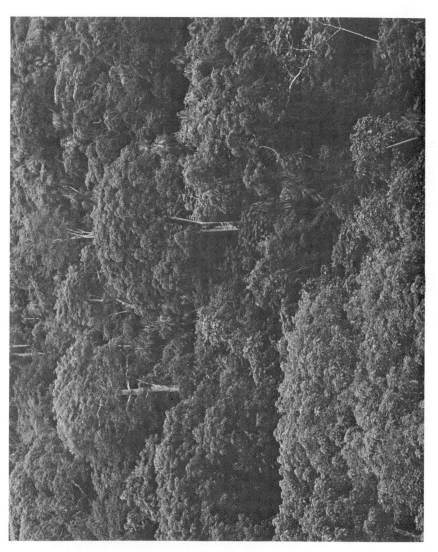

Figure 2.2.    Aerial view of Tropical Moist forest in eastern Panama. (Photograph by Peter McGrath.)

Figure 2.3.   A view of the Tropical Moist forest during the dry season. The trees without leaves are emergent cuipo. (Photograph by G. Child at Santa Fe, February, 1967.)

successional are the forests situated in Darien between the River Tuira and the present Panama Canal. These forests were originally classed as primary and considered as semidry giving the deciduous character to many of the dominants. However, a more detailed knowledge of the successional processes shows that the dominants are, in reality, secondary species that are not reproducing but attain great size. Under the climate of this region the dominants of the primary forest will not be deciduous, but perennial. Only an intense occupation four centuries ago and a subsequent abandonment offers an explanation of this situation." This conclusion of Budowski is partly based on the historical evidence of early Spanish explorers— "Balboa reported [letter of January, 1513] that to the north at a day's journey from the caicque Pocorosa are some sierras the most beautiful that have been seen in these parts. They are very open *(claras)* mountains without any woods *(monte),* save for groves *(arboleda)* along some arroyos that come down from the mountains" (Sauer, 1966 p. 285). Sauer goes on to state that "by the early accounts the sigmoid lowland that runs the length of the isthmus was more or less continuous savannah." The evidence seems to show that the forests studied in Darien, Panama, are about four hundred years old and have replaced the open savannah lands maintained by Precolombian indians. The question of the character of the primary forest remains unanswered. Here we have accepted the forest as it exists, assuming that it responds to the strong dry season even in the primary state. In this report we have called this forest Tropical Moist forest after the Holdridge system, recognizing that, actually, it might be transitional between Tropical Dry and Moist forests. One object of this section is to provide the reader a description of the forest sufficiently adequate to enable him to place it in whatever classification system is most useful and meaningful to him.

## THE ENVIRONMENT

The Darien Province experiences a well marked dry season from January to April in most years. During 1967, January, February, and March were arid months (fig. 2.4). Rain falls during eight to nine months (May to December), for a total annual rainfall of approximately 200 centimeters. The temperature remains relatively constant throughout the year, with a mean annual temperature of about 25°C.

The parent material of the soils in the area of interest is a shale, which contains interbeds of dolomite and calcareous sandstone (IOCS-FD-31, 1967). The beds of shale are described as Sabana Beds of

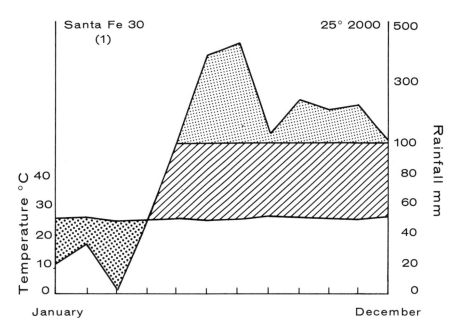

Figure 2.4. Climate diagram (following the form of Walter and Lieth, 1960-1967) for Santa Fe, showing temperature and rainfall over one year. Note the dry season in January, February, and March.

the Upper Miocene age (IOCS-FD-31, 1967). Petrographic analysis of Rio Lara calcareous shale showed that it is made up of 65 percent clay (montmorillonite or illite), 25 percent calcite quartz, and 15 percent feldspar, collophane, and organic matter (IOCS-FD-31, 1967). The soils are almost all residual from these underlying rocks and average, in the shale of the Sabana Beds, weathering depths of 8 to 14 meters.

The surface soil of the Tropical Moist forest to a depth of 30 cm. appears dark grey or black and seems to be similar to the margalitic soils described by Mohr and Van Baren (1954) and dark clay soils of Dudal (1963) for other tropical areas with a distinct dry period. Montmorillonite is the major clay mineral in these margalitic soils, lime concretions occur in the profile, and calcium and magnesium make up 90 percent of the absorbed cations. Dudal (1963) states that "a common characteristic of the climatic conditions in which dark clays occur is a very marked seasonal distribution of the rains and the

occurrence of a distinct dry season. The climate thus permits a sufficiently rapid weathering of the rocks to take place but does not cause
a continuous and rapid leaching of soluble elements. This process is
strengthened by the slow permeability of the heavily-textured materials." Under the changing moisture conditions these soils show
very strong swelling and severe shrinking. Gamble et al. (1969),
studying the clay fractions of soils in the Tropical Moist forest
in the Darien Province, showed by X-ray diffraction that the clay
was interlayered montmorillonite—halloysite and has a surface area
of 575 to 1030 square meters per gram. We also observed concretions
in the profile and soil analyses showed that calcium and magnesium
were the major cations. Further, our limited soil analyses (four mixed
samples from the upper 30 centimeters from one site) indicated that
this soil has an exchange capacity of about 50 milliequivalents per
100 grams and is about 86 percent base saturated. These data suggest
that the soil is relatively youthful (Sherman, 1971).

THE FOREST

The comments above briefly describe the environment of the Tropical Moist forest. Next, our object is to define the forest in terms of
its distribution, shape and mass, and to identify the requisite ecosystem compartments. The reader will appreciate the problem of
mapping vegetation over a large region of difficult terrain, where the
sole method of surface transport is the dugout canoe and where
aerial photography is inadequate. Since it was not essential that we
develop a detailed geographical map, our method was to fly transects
across the area by airplane or helicopter and map the appearance
of emergent cuipo or bongo above the canopy. Prior ground reconnaissance and discussions with inhabitants indicated that the presence of these two trees was indicative of Tropical Moist forest. Using
this technique we determined that Tropical Moist forest covers about
75 percent of the land area in the region of concern (fig. 2.5); that is,
practically all of the land below an elevation of 250 meters which is
not periodically inundated. At higher elevations the forest was clas-
sified as Premontane Wet forest. Where land was periodically
flooded, forest was either Riverine or Mangrove forest.

Since the Tropical Moist forest is characterized by the response of
the emergent trees to a wet-dry regime it was desirable to sample it
during both seasons; this was accomplished in February and September (See fig. 2.4 for rainfall during these months). The actual sampling sites were dictated by support facilities. Since the major inter-

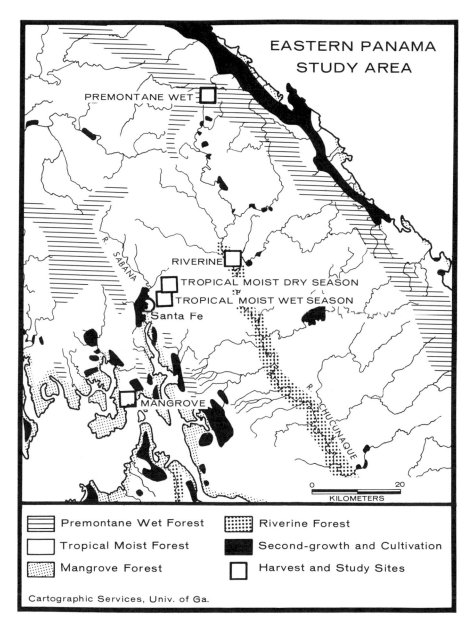

**EASTERN PANAMA STUDY AREA**

PREMONTANE WET

RIVERINE

TROPICAL MOIST DRY SEASON

TROPICAL MOIST WET SEASON

Santa Fe

MANGROVE

R. SABANA

R. CHUCUNAQUE

0    20
KILOMETERS

| | Premontane Wet Forest | | Riverine Forest |
| --- | --- | --- | --- |
| | Tropical Moist Forest | | Second-growth and Cultivation |
| | Mangrove Forest | | Harvest and Study Sites |

Cartographic Services, Univ. of Ga.

Figure 2.5. Detailed map of the study area, showing the region lying between the lines on figure 2.1. The majority of the area is classed as Tropical Moist forest (open). The Premontane Wet forest (heavy lines) appears on the highlands, Riverine forest (heavy dotted) occurs along the Chucunaque River, and Mangrove forest (light dotted) occurs on the coast line. The locations of the sampling areas for each type are shown as open squares.

oceanic canal base camp was located at Santa Fe (fig. 2.5) in the Tropical Moist forest, it was convenient to locate our two study areas near this camp. The dry season site was about five miles from Santa Fe on the straight-line road cut part way across the isthmus by the engineers and was on the Rio Sabana drainage. The wet season site was in the opposite direction, within walking distance of Santa Fe on the Rio Lara drainage since the cross isthmus road was impassable at this period of the year. In both cases the sites were chosen after a reconnaissance of the area on foot showed that they most nearly represented the typical condition of the surrounding forest based on their topography, number of tree strata, and presence of conspicuous species. We assumed in selecting these sites that they sampled the same habitat. The technical locations of these two study areas are: wet season, Rio Lara site, latitude 8° 38′, longitude 78° 08′; dry season, Rio Sabana site, latitude 8° 41′, longitude 78° 07′.

We have mentioned that sampling on these sites was by one-quarter-hectare plot. It is worth examining the basis for choosing this size of plot since plot size strongly influences the results of vegetation sampling. The conventional way to choose a plot is to compare characteristics of the vegetation, such as the number of species, or the weight of plants, on plots of increasing size. The minimal sampling area is indicated when an increase in plot dimensions does not result in a change in the parameter of interest (Cain and Castro, 1959). For example, in considering numbers of plant species a point is reached in the sampling program where few or no new species are encountered by increasing the size of the area sampled. This size plot then is the most efficient for determining the number of species.

Several authors have been interested in the problem of determining the numbers of species of plants in the tropical forest. Drees (1954) counting all plants of all sizes found a plot of one-quarter hectare was adequate to determine species numbers in Indonesia. However, if he considered only trees with a diameter over 30 centimeters, then a plot of one hectare was not sufficient to sample the species. Richards (1939) found the latter also to be true in south Nigeria in his examination of trees over 10 centimeters in diameter. Pires et al. (1953) found 3.5 hectares was not large enough to sample trees over 10 centimeter diameter in Amazon forest. Indeed, Cain et al. (1956) showed that the species-area curves for trees over 10 centimeter diameter in Brazilian rain forest continually increase to 20,000 square meters, their largest plot size. Apparently the minimum plot size to sample the number of species in tropical forest depends on

Figure 2.6.   Interior view of the Tropical Moist forest. Note large bole of an emergent cuipo (*Cavanillesia plantanifolia*) and the guagara palm (*Sabal allenii*). (Photograph by the authors.)

whether the researcher is interested in all individual plants or only the larger trees. As far as we know there are no analyses of plot size and forest mass, the parameter we wished to sample, in the literature. Since we were interested in the mass of all species and sizes of plants we chose Drees's plot of one-quarter hectare for our study. Further, Japanese ecologists working in Thailand forests (Tsutsumi et al., 1967) also used a quarter-hectare plot and, thus, our data can be directly compared with theirs. Finally, the quarter-hectare plot fit most closely to the limits of our supply of labor and equipment. Further study will be required to determine if a quarter-hectare plot gives the best representation of the mineral and biomass content of the Tropical Moist forest.

ABUNDANCE AND DISTRIBUTION OF STEMS

Visitors to a tropical forest are often struck by the size, number, and height of the trees. The quotation from Wallace, which heads this section, expresses this general feeling. We were equally impressed by the Tropical Moist forest (fig. 2.6). In the Darien the maximum tree height of this forest is about 40 meters and the canopy is broken into several strata (fig. 2.7), with the palm, *Sabal allenii*, occuring frequently in the subcanopy. The stems of trees (over 10 centimeter DBH) are not evenly distributed throughout the quarter hectare (fig. 2.8). A *chi*-square test of goodness of fit to the Poisson distribution (Greig-Smith, 1964) also showed that the stems were not distributed at random ($P > 0 \cdot 10$). The apparent aggregation of larger stems probably reflects small differences in relief, but these were not studied.

The number of stems on both plots was about 4,800 (table 2.1), although the numbers of individual palms, trees other than palms,

Table 2.1.   Number of plant forms on one-quarter hectare
in Tropical Moist forest in Panama.

| Month sampled | Trees other than palms | Palms | Fleshy plants | Vines | Total |
|---|---|---|---|---|---|
| February | 1087 | 63 | * | 3633 | 4783 |
| September | 1512 | 142 | 244 | 2856 | 4754 |

* Not measured separately because of small numbers present.

Figure 2.7. Cross-section drawing of Tropical Moist forest showing the strata of the canopy. (Drawing by P. M. Golley from a photograph.)

# Tropical Moist Forest

## Subplot 1

## Subplot 2

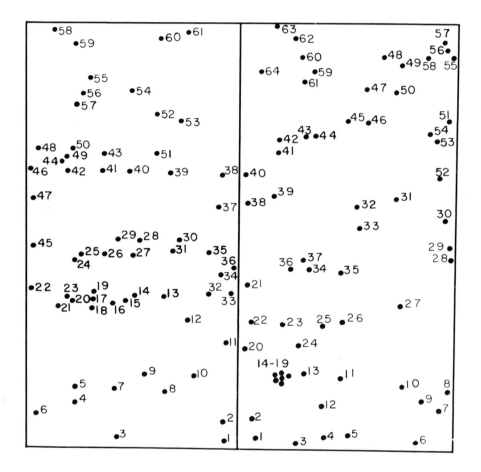

Figure 2.8. Distribution of stems in Tropical Moist forest. The points on the map represent trees over 10 cm DBH at the Rio Lara site. See Appendix, table 6, for diameter and names of trees corresponding to the numbers on the map.

and vines differed between the sites. More vines were encountered on the site sampled in February and more trees and palms occurred on the plot harvested in September. Many non-woody plants, such as *Heliconia,* also were encountered on the Rio Sabana plot.

On both plots there were a few very large stems with a DBH greater than 100 centimeters, but most stems were of the smallest diameters (see fig. 1.5). Many of the smallest stems were of vines which were abundant throughout all parts of both plots. The average DBH of all stems was 1.60 ± 0.06 centimeters on the Rio Sabana plot and 1.77 ± 0.08 centimeters on the Rio Lara plot.

For trees alone on the Rio Sabana and Rio Lara plots, discounting vines, palms, and fleshy plants, the average DBH was 4.18 ± 0.22 and 3.17 ± 0.22 centimeters respectively. The largest trees were roble *(Tabebuia pentaphylla)* and cuipo. The DBH, Spanish name, and scientific name, when known, for the stems encountered on the Rio Lara plot are listed in appendix table 6.

OTHER DESCRIPTIVE FEATURES OF THE FOREST

Another forest characteristic which was measured for definition purposes but does not relate directly to the mineral cycling study, is the size and area of the tree canopy. The percent of the total canopy area covered by leaves was estimated by taking 200 readings, using a sighting tube 27.5 centimeters long by 3.5 centimeters wide held at eye level, looking directly above at the canopy. The readings were taken at 10 meter intervals across the study plot and in the surrounding forest. While the dry season site had a greater percentage cover (84%) than the wet season plot (78%), these differences are not statistically significant at the 0.5 level of probability.

The leaf surface (leaf area index, LAI) was determined by selecting 100 leaves at random from the piles of harvested leaves, weighing them, and outlining each on a piece of graph paper. The paper outlines were cut and weighed and the weights converted to surface area. The average surface area per gram of leaf was multiplied by the total weight of leaves on the plot to estimate the LAI of the forest. Both sites had a high LAI. However, the index for the dry season site $(10.6 \ m^2/m^2)$ was only one-half that of the wet season site $(22.4 \ m^2/m^2)$. A wet-dry season difference is expected since the emergent trees lose their leaves during the dry season, but the value for the wet season plot is very high. For example, Kira et al. (1964) report a LAI of $12 \ m^2/m^2$ for rain forest in Thailand, and Odum (1970) reports $7.34 \ m^2/m^2$ for Montane forest in Puerto Rico. There have been high values of LAI re-

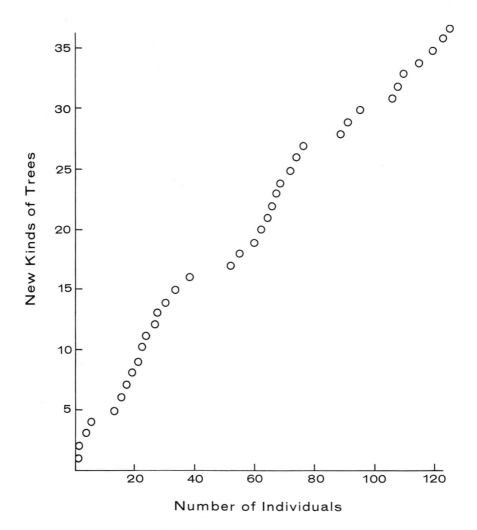

Figure 2.9. The relationship between the number of individuals and new kinds of trees with a DBH over 10 cm on a one-quarter hectare plot of Tropical Moist forest at Rio Lara, Panama. The kinds of trees were identified by Indian helpers and are listed under their Spanish names in appendix table 6.

ported for other vegetation; Ogawa et al. (1961) reports a LAI of 23.3 for an *Arundo donax* community in Japan and Burger (in Tadaki, 1966) reported 28 for *Picae abies* and 27.1 for *Pseudotsuga douglasii.* Even so, the leaf area of the Tropical Moist forest is unusually large.

We did not attempt to make a species list for the Tropical Moist forest but our colleague, James Duke, Battelle Memorial Institute, compiled a list of species for us which is derived from his, as yet, incomplete studies of the plant distribution and ecology of the region. The list (appendix table 4) of identified specimens contains 395 taxa, of which 35 are understory herbs, 30 are understory palmitas, 56 are understory shrubs, 42 are woody vines, 15 are herbaceous vines, 11 are epiphytes, 157 are subcanopy trees, and 49 are canopy trees. While this is not a complete count of species, it indicates a rich tropical flora. In comparison, Pires et al. (1953) concluded that the number of tree species in the Amazon forest was near 250. In another area of the Amazon, Cain et al. (1956) calculated that there were about 200 species of trees with a DBH over 10 centimeters. Richards (1945) states that there are estimated to be 596 species of large trees in the whole of the Ivory Coast. Duke's incomplete list suggests that the number of species in the Tropical Moist forest of Darien probably is similar to these other tropical forests.

Finally, an analysis of the relationship between the number of trees identified by our Indian coworkers and the number of individual trees over 10 cm. DBH on the plot sampled in September (fig. 2.8, and appendix table 6) reflects the diversity of the forest. Presumedly the Spanish names of trees are roughly comparable to the species names. We encountered 37 types of trees on the one-quarter hectare plot, while Duke's preliminary list of species included 49 canopy trees and 157 subcanopy trees in the forest type. The number of new species increases regularly with the number of individuals up to the 125 individuals over 10 cm DBH counted on the plot (fig. 2.9). Clearly the one-quarter-hectare plot size was inadequate to sample the taxonomic variety of the larger trees in the Tropical Moist forest, as was mentioned earlier.

# CHAPTER 2

# The Forest Biomass

The descriptive data presented in chapter 1 are mainly useful in the definition of the Tropical Moist forest ecosystem; they are not directly applicable to the description of mineral cycling in the forest. In contrast, the weight or biomass of the forest is fundamental to a mineral cycling study. The biomass represents stored organic matter in the ecosystem; it specifies the numerical value of the ecosystem components. With information of the organic biomass and the concentration of elements in that biomass we can calculate the chemical inventory of the forest. The living biomass consists mainly of water, the amount of which varies with the compartment and season. When the living tissue is dried and the water driven off the remaining material is composed mainly of carbohydrates, such as cellulose and lignin, and a smaller amount of protein. If this dry tissue is burned to drive off the carbon, hydrogen, oxygen, together with the nitrogen and sulfur from the protein, an ash remains which contains the mineral elements. According to Rodin and Bazilevich (1967) these mineral elements in tropical vegetation make up about four percent of the dry weight biomass. While this quantity of minerals is quite small, it is essential for the proper functioning of the system, and a lack of adequate minerals can limit growth and development of the forest even if other limiting factors are available in excess.

## PROCEDURES

Since information on the biomass is basic to the mineral cycling study we will describe the procedures for its measurement in detail. We have already alluded to the general sampling approach in the introductory section. Our objective was to establish the wet and dry weights of the five vegetation compartments of the ecosystem: the leaves, stems, fruits and flowers, litter, and roots. This was accomplished by harvesting the vegetation on quarter-hectare plots.

The harvest program proceeded as follows. First, litter was sampled in the undisturbed forest plot on ten square-meter subplots located at random. Litter included leaves, twigs, stem material, fruits, and flowers. Next, all understory vegetation to about two meters height was cut at ground level. Leaves were stripped from stems (fig. 2.10) and the fruits, flowers, and stems were collected separately. These materials were immediately weighed on a beam balance (fig. 2.11). For convenience these understory materials are reported separately from overstory.

Next, samples of the overstory were taken from the distribution of tree diameters. All trees with a DBH under 10 centimeters were harvested and the weight of their stems, leaves, and fruits determined. Within seven size classes over 10 centimeters DBH, 10-15, 15-20, 20-30, 30-50, 50-100, 100-200, and over 200 centimeters DBH, at least 10 percent of the trees and palms, with their associated vines and epiphytes, were harvested. The largest trees could not be cut and weighed entirely and subsampling was necessary. The diameter and length of the tree bole was measured and its volume calculated as:

$$\text{Volume} = \frac{\pi h}{3} (r_1^2 + r_1 r_2 + r_2^2)$$

where $h$ is the height, $r_1$ is the radius at the base of the log and $r_2$ is the radius at the top of the log. A known volume of log was weighed and the volume of the entire log was converted to weight with this information. On these larger trees the branches were counted and one half or more were cut, stripped of leaves and the wood, leaves, and fruit were weighed. In this way the total weight of the above-ground portion of the large tree could be reconstructed.

To complete the calculation of forest biomass, it was necessary to convert the number of stems per diameter class into accumulated biomass for the forest. First, a relationship between stem diameter and leaf or stem biomass was established. The total weight measured in the field, corrected for moisture content, was divided by the number of stems and expressed as the mean weight of an individual tree of a diameter at the midpoint of that diameter class. For example, five trees weighing 40 kilograms, in the diameter class 10 to 15 centimeters, were expressed as an individual of 8 kilograms, with a diameter of 12.5 centimeters. Richard Clements made a number of tests of goodness of fit between these data on weight and diameter and found that the relation between the cube root of biomass of stems or leaves and DBH gave the highest correlation. The coefficient of determination ($R^2$) for stems was 0.986 and 0.887 for the Rio Sabana plot and Rio Lara plot and for leaves was 0.923 and 0.762 respectively.

Figure 2.10.    Stripping leaves from stems in the forest. (Photograph
by the authors.)

Figure 2.11. Weighing leaf material in the forest. The scientist in the photograph is Jack Ewel (University of Florida), who has provided data on second-growth biomass. (Photograph by the authors.)

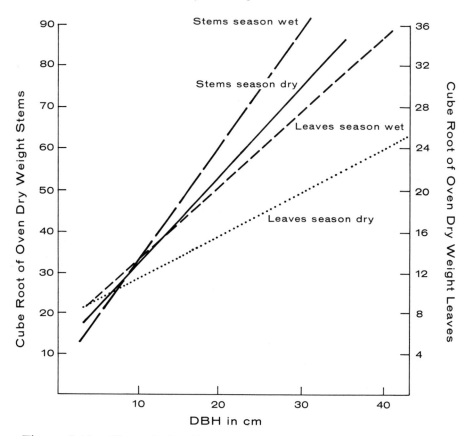

Figure 2.12. The relationship between leaf and stem biomass and tree diameter. The graphs show the regression of the cube root of oven dry weight biomass on tree diameter (DBH) in dry season and wet season Tropical Moist forest.

Using these regressions, the formulas for leaves were:

$$\text{Rio Sabana} \quad \sqrt[3]{\hat{Y}} = 7.14 + 0.5081X$$
$$\text{Rio Lara} \quad \sqrt[3]{\hat{Y}} = 5.82 + 0.7112X$$

and for stems,

$$\text{Rio Sabana} \quad \sqrt[3]{\hat{Y}} = 10.82 + 2.1093X$$
$$\text{Rio Lara} \quad \sqrt[3]{\hat{Y}} = 4.28 + 2.7584X$$

where $\hat{Y}$ is the cube root of leaf or stem biomass and $X$ is the DBH of the trees. The regressions lines (fig. 2.12) differ for the two seasons.

As noted before, the wet season forest had a larger mass of leaves in relation to stem diameter, but this difference would be expected since the emergent trees had lost their leaves in the dry season. However, the wet season plot had a higher weight of stems per diameter indicating a basic difference between the plots, which would also exaggerate the difference in leaf biomass.

With these regressions a computer program was written which multiplied the mean biomass by the number of stems to give the weight for that diameter class. Finally, each class was summed for the total weight of leaves and stems per plot.

The remaining vegetation compartment, the roots, was sampled on ten square-meter plots. The soil was loosened with a pick and shovel and roots were sampled to a depth of 0.3 meters which appeared to be the depth of the major portion of the root mass. Roots of a size greater than 0.05 millimeters diameter were collected and weighed.

Epiphytes were collected separately, weighed, and then later added to leaf or stem compartments as appropriate.

## THE WATER CONTENT OF THE BIOMASS

The water content as a percentage of the total weight is shown in table 2.2. These percentages were calculated as the ratio of material dried at $100°C$. to constant weight to the weight of freshly-cut tissue. In almost every compartment the water content during the wet

Table 2.2.    Percent water content of Tropical Moist
forest vegetation. Values represent averages with one
standard error in parentheses.

| Compartment | Dry season February | Wet season September |
|---|---|---|
| Overstory leaves | 63.2 (3.2) | 69.9 (1.5) |
| Understory leaves | 64.3 (2.5) | 73.8 (1.3) |
| Overstory stems | 48.4 (0.4) | 66.7 (7.9) |
| Understory stems | 54.2 (0.3) | 65.3 (1.7) |
| Overstory fruits and flowers | 88.8 (2.4) | 76.8 (4.9) |
| Understory fruits and flowers | 80.1 (8.4) | 85.4 (2.0) |
| Litter | 20.9 (0.5) | 72.4 (0.6) |
| Roots | 60.3 (1.7) | 61.8 (2.9) |

Table 2.3.    Biomass of Tropical Moist forest
in Panama (kg dry weight/ ha).

| Compartment | Rio Lara | Rio Sabana |
|---|---|---|
| Overstory leaves | 11369 | 7349 |
| Overstory stems | 354735 | 252126 |
| Understory leaves | 620 | 744 |
| Understory stems | 1094 | 3271 |
| Fruits and flowers | 139 | 8 |
| Roots | 9850 | 12633 |
| Total living biomass | 377807 | 276131 |
| Litter, excluding dead wood | 2910 | 6200 |
| Dead wood | 14644 | * |
| Total dead biomass | 17554 | 6200 |

* Not measured.

season was greater than during the dry season. The average percentage of water in the biomass was 65.7 percent in the wet season and 51.8 percent in the dry season. These averages differed significantly at 0.5 level of probability, by the Duncan's New Multiple Range Test of differences. The litter compartment reflected the greatest seasonal difference in water content. Fruits and leaves contained the greatest quantities of water for any compartment in both seasons. Water content of the Tropical Moist forest wet season compartments was similar to those of the Montane forest at El Verde, Puerto Rico (Odum, 1970). Odum's estimates include: leaves, 68 percent moisture; boles, 52 percent; branches, 59 percent; fruit of dicotyledons, 77 percent; roots, 41 percent; and leaf litter, 76 percent.

FOREST BIOMASS

The organic mass of the forest was greatest at the Rio Lara site (table 2.3). The observed differences are mainly a function of location since the wet season site also had a greater density of trees and a larger basal area of stems (Rio Lara basal area of 11.3 $m^2$/quarter hectare compared to Rio Sabana basal area of 6.6 $m^2$/quarter hectare). Probably the only strictly seasonal difference is the quantity of litter, excluding deadwood, which was greater at the dry season site. Since the dry season is the time of leaf fall this difference is expected.

Table 2.4.   Biomass of selected tropical forest communities.
(Metric tons dry weight per hectare.)

| Forest | Canopy leaves | Canopy stems | Understory | Roots | Litter |
|---|---|---|---|---|---|
| Thailand Rain forest[1] | 8.2 | 360 | 2.4 | 33 | 3.5 |
| Thailand Monsoon forest[1] | 3.8 | 261 | 2.0 | 25 | —— |
| Puerto Rico Lower Montane forest[2] | 8.1 | 269 | —— | 71 | 19.3 |
| Congo Secondary forest[3] | 6.5 | 116 | —— | 31 | 5.6 |
| Tropical Moist | | | | | |
|    Rio Sabana | 7.3 | 252 | 3.9 | 13 | 6.2 |
|    Rio Lara | 11.3 | 355 | 1.7 | 10 | 2.9 |

[1]Ogawa et al. (1965b).
[2]Odum et al. (1970).
[3]Greenland and Kowal (1960).

Comparison of the Tropical Moist forest with other forests is re-
served for section 4 but it is important at this point in the analysis
to know if the biomass estimates are reasonable. Comparison of the
Tropical Moist forest biomass with that of other tropical forests
(table 2.4) indicates that the Tropical Moist forest does not exhibit
unusual weights of leaves, stems, understory, or litter. However, the
estimates of root biomass appear to be low. A regression of root
biomass on stem biomass, using extensive data from forests in Ovington (1965), showed that roots are usually about 24 percent of the
total stem weight. If this relationship is true for the Tropical Moist
forest, then total root biomass should be 24 percent of the stem
weights in table 2.3. Apparently, we sampled only about 20 percent
of the estimated root biomass with our pits of 0.3 meter depth. However, the underestimate is not solely due to depth of sampling since
Greenland and Kowal (1960) found in an African tropical forest
that 86 percent of the total root mass present to a 1.5 meter depth
occured in the first 30 centimeters and Odum (1970b) shows similar
root profiles for Tropical Moist and other forests throughout the
Caribbean. Probably our underestimate was due to missing the tap
roots and the root crowns directly beneath the stumps.

## THE BIOMASS OF ANIMALS

Logically the biomass of animals is handled in the same way as the
plants. Four kinds of information are needed to estimate the chemical

content of the animal compartments of the ecosystem—the density of populations, the weight of individuals, the feeding habits of individuals, and their elemental concentration. Unfortunately the density and feeding habits of animal populations often are exceedingly difficult to measure. Each species population has unique habits which prevent the use of a single census technique for the numerous populations. Further, the animals are distributed in complex patterns throughout the strata of tall forest (Allee, 1926). Food habit determination requires close observation in the field, examination of stomachs, or experimental techniques. For some forms neither census nor feed analysis methods are available.

In this situation our approach was to collect specimens of all trophic groups of animals for weight and chemical analysis. These collections were made in a systematic way by sweep-netting insects and invertebrates, mist-netting birds, and randomly hunting reptiles, amphibians and mammals. In most cases, the food habits and the density of these organisms were derived from the literature. With these data we estimated forest animal biomass.

The animal populations were considered first in groupings of mammals, birds, reptiles and amphibians, invertebrates living above ground, ground macroarthropods and ground microarthropods since density and other data in the literature are usually reported in this way. Later each of these groups was divided into herbivores, carnivores, and detritus feeders.

Insect populations were determined by sweep-netting the understory vegetation. Twelve samples of 100 sweeps using a 37.5 centimeter diameter net were made in each forest site. Diptera, Hymenoptera, Araneae, and Coleoptera were collected most frequently by sweep-netting. The 100 sweeps collected an average of 55 insects in the dry season forest and 109 insects in wet season forest (table 2.5) and these represented a live weight of about 0.24 and 0.36 grams respectively. Presumably the large number of insects reflects the greater quantity of fruits and flowers and the more succulent vegetation available in the wet season (table 2.4). The dry weight of insects was determined by drying the samples at 100° C. for 24 hours. The percent water in insects after preservation in vials of alcohol was 23 percent.

To convert these data to density and biomass we used the assumption of Crossley and Howden (1961) and Gray and Treloar (1933), that 10 sweeps of the insect net collect the population on one square meter of vegetation up to 2 meters height. Next, we expanded

Table 2.5.    Numbers and live weight (grams) of insects
collected in Tropical Moist forest by sweep nets.

| Sample | Dry season | | Wet season | |
|---|---|---|---|---|
| | Number/100 sweeps | Live weight | Number/100 sweeps | Live weight |
| 1 | 30 | * | 115 | 0.359 |
| 2 | 24 | * | 80 | 0.226 |
| 3 | 71 | * | 105 | 0.235 |
| 4 | 47 | * | 79 | 0.100 |
| 5 | 94 | * | 41 | 0.032 |
| 6 | 40 | * | 101 | 0.232 |
| 7 | 38 | * | 116 | 1.720 |
| 8 | 104 | * | 136 | 0.227 |
| 9 | 47 | * | 142 | 0.363 |
| 10 | 60 | 0.245 | 91 | 0.118 |
| 11 | 75 | * | 156 | 0.357 |
| 12 | 30 | * | 143 | 0.420 |
| Total | 660 | | 1305 | |
| Average | 55 | 0.245 | 109 | 0.366 |

* refers to samples dried out and not used.

this estimate to the entire column of vegetation of 40 meters height. In the absence of any information on the stratification of the fauna with height, we have assumed that the numbers of insects are constant throughout the vegetation column. This may be a false assumption, since the midpoint of the vegetation column has few leaves while the canopy has many leaves and a large insect fauna, but we do not know what difference we should expect with height. With these assumptions, we calculated that the average biomass of insects as the live weight per 100 sweeps sampling two meters of understory times 20 equalled the weight of insects in a volume of vegetation of 10 square meters area and 40 meters height. This value multiplied by 77 percent dry matter and divided by 10 gave the seasonal dry weight per square meter estimates of insect biomass. The dry season insect biomass ($0.38\ g/m^2$) was less than the wet season ($0.56\ g/m^2$). The wet season estimate was used in further calculations of animal biomass since it represents the population over the largest portion of the year.

Table 2.6.    Distribution of animal biomass in a Panama forest.

|  | Dry weight $g/m^2$ | |
|---|---|---|
| Herbivores | | |
| Mammals | 0.71 | |
| Arthropods | 0.49 | |
| Total | | 1.20 |
| Frugivores | | |
| Mammals | 0.08 | |
| Birds | 0.01 | |
| Arthropods | 0.01 | |
| Total | | 0.10 |
| Carnivores | | |
| Mammals | 0.04 | |
| Birds | 0.02 | |
| Reptiles and amphibians | 0.71 | |
| Arthropods | 0.06 | |
| Total | | 0.83 |
| Detritivores | | |
| Soil and litter macroarthropods | 1.88 | |
| Soil and litter microarthropods | 3.36 | |
| Total | | 5.24 |
| Total animal biomass | | 7.37 |

This estimate of insect biomass can be compared to estimates of Janzen and Schoener (1968) for the understory of four forests in Costa Rica. The number of individuals collected by sweep-netting there averaged 2,254, and ranged from 653 to 5,109 per 2,000 sweeps. The weight of insects averaged 3,955 milligrams per 2,000 sweeps and ranged from 2,661 to 5,344 milligrams. Assuming that 10 sweeps sampled a square meter and that they sampled $1/32$ of the vegetation column (0.5 to 1.75 meters above ground), then their collections represent 0.43 to 0.86 grams per square meter of ground surface. This estimate is very similar to our estimate for the Tropical Moist forest.

The arthropod fauna was divided into trophic groups assigning 10 percent to the carnivores, 2 percent to fruit eaters and 88 percent to herbivores (table 2.6). This separation was made on the assumption that 10 percent of the herbivore mass is transferred to carnivores and that the herbivores feed on leaves and fruit in proportion to the biomass of these foods.

The above ground vertebrates were less easily sampled and it was necessary to use literature data entirely to estimate their biomass. Harrison (1969) has reported densities and biomass of mammals in lowland Tropical forest in Malaya. In a 14.6 hectare area of disturbed lowland forest he determined 35 species were present at a density of 7.2 per hectare. Trapping on the area resulted in capture of 4.9 individuals per hectare with a biomass of 171 grams per individual. We estimate the other 2.3 animals per hectare, which represent the larger forms such as leopard, deer, and monkeys listed by Harrison, weighed an average of 25 kilograms live weight or 7,500 grams dry weight. The total mammal biomass then was 840 plus 7,500 grams or 8,340 grams per hectare (0.83 grams per square meter). The mammal biomass was distributed among the trophic levels according to the known feeding habits of the animals listed by Harrison (table 2.6). Most of the mammal biomass is herbivore.

Mist nets yielded approximately equal numbers of birds per net day (table 2.7) for three different Darien forests, including the two Tropical Moist forest sites. However, more species (32) were captured in the Tropical Moist wet season forest than in the other locations. The average weight of the birds was 37 grams live weight (table 2.8). These mist net data could not be converted directly to density estimates since we do not know the area sampled by the nets. We averaged the data from 57 censuses of bird populations in the tropics for a density estimation of 24.5 per hectare (991 per 100 acres). The estimated density was divided into trophic groups on the basis of data from Harrison (1962), and multiplied by the average biomass of an individual bird from the mist net captures, and finally, converted to dry weight (assuming 70 percent of the live weight as water). This calculation gave an average bird biomass of 0.027 grams per square meter. About 75% of the bird mass was assumed to be in the carnivore group and 25% to be fruit or seed feeders (table 2.6). Karr (1971) has recently reported on bird populations from a moist lowland forest from a similar area in Panama. He found a similar average individual weight for birds, a slightly higher standing crop (0.040 $g/m^2$) and gave a similar breakdown of food habits of the species population.

Systematic collections of reptiles were not made but excellent population data are available for these animals in Panama from Heatwole and Sexton (1966). Their population densities of animals on the forest floor and low vegetation (0.47 per square meter) multiplied by a weight factor of 1.5 gram dry weight per individual,

Table 2.7.    Birds collected by mist nets in three forests
in Darien Province, Panama (names from Eisenmann, 1955).

| | Tropical Moist (wet season) | Premontane Wet | Tropical Moist (dry season) | Riverine |
|---|---|---|---|---|
| Number of days | 7 | 5 | 2 | 2 |
| Number of nets | 12 | 10 | 11 | 10 |
| **Species** | | | | |
| Plainenops | 2 | | | |
| Scaly-throated leafscraper | 1 | 1 | | |
| Black-bellied wren | 2 | | | |
| Golden-collared manakin | 1 | | | |
| Green manakin | 2 | | | |
| Olivaceous woodcreeper | 1 | | | |
| Black-faced antthrush | 3 | | | 1 |
| Fly catcher | 2 | | | |
| Buoy-tailed hermit | 6 | | | |
| Gray-headed tanager | 2 | | 4 | |
| White-breasted wood wren | 1 | 1 | | |
| Spotted antbird | 6 | | 5 | |
| Bare-crowned antbird | 1 | | 1 | |
| Slaty ant shrike | 5 | 1 | 2 | |
| Olivaceous flatbill | 3 | | | |
| White-bellied antbird | 3 | | | |
| Hummer | 1 | | | |
| White-whiskered puff bird | 1 | | 1 | 2 |
| Sulphur-rumped flycatcher | 2 | 3 | | 1 |
| Long billed gnat wren | 1 | | | |
| Kentucky warbler | 1 | | | |
| Buff-throated woodcreeper | 2 | | | |
| White-flanked antwren | 1 | | 2 | |
| Checker-throated antwren | 2 | | 2 | |
| Immaculate antbird | 1 | | | |
| Bicolored antbird | 7 | 1 | 1 | |
| Ocellated antbird | 5 | | 1 | |
| Blacktailed flycatcher | 1 | | | |
| Chestnut backed antbird | 2 | | | |
| Great ant shrike | 1 | | | |
| Striped wood haunter | 1 | | | |
| Bandtailed barbthroat | 6 | | | |
| White-ruffed manakin | | 15 | | |
| Golden-crowned spadebill | | 1 | | 1 |

Table 2.7.    Birds collected by mist nets in three forests
in Darien Province, Panama (names from Eisenmann, 1955), *cont.*

| | Tropical Moist (wet season) | Premontane Wet | Tropical Moist (dry season | Riverine |
|---|---|---|---|---|
| **Species** | | | | |
| Olive-striped flycatcher | | 4 | | |
| Tanager | | 14 | | |
| Wedgebilled woodcreeper | | 11 | | 2 |
| Streaked chested antpitta | | 1 | | |
| Unidentified hummingbird | | 3 | | |
| Sicklebill | | 1 | | |
| Spotted woodcreeper | | 1 | | |
| Broadwing hawk | | 1 | | |
| Wing-banded antthrush | | 1 | | |
| Half-collared gnatwren | | 1 | | |
| Unknown | | 3 | | |
| Long-tailed hermit | | | 1 | 3 |
| Keel-billed toucan | | | 1 | |
| Crimson-crested woodpecker | | | 1 | |
| Long-tailed woodcreeper | | | 2 | |
| Swainson's thrush | | | 1 | |
| Summer tanager | | | 1 | |
| Golden-headed manakin | | | | 6 |
| Checker-throated antbird | | | | 1 |
| Ruddy tailed flycatcher | | | | 3 |
| Barred woodcreeper | | | | 1 |
| Total | 76 | 64 | 26 | 21 |
| Captures per net day | 0.91 | 1.28 | 1.18 | 1.05 |

based on unpublished data of Michael Duever, results in an average reptile-amphibian biomass of 0.71 grams per square meter. All reptiles and amphibians were considered carnivores.

Our sampling methods did not include the ground or soil dwelling organisms and we have entirely relied on literature data. Several reports on the soil fauna are available from tropical regions. Strickland (1947) presents data from cacao plantations and savannah habitats in Trinidad, Wiegert (1968) reports populations from Puerto Rico Montane forest, and Madge (1965) has data from forest litter in

Nigeria. Strickland found 237 invertebrates in cores 3.6 inch diameter by 3 inch deep in cacao and 90 and 255 in cores from savannah. Wiegert reported macroarthropod populations ranging from 2,750 to 5,590 individuals per square meter in litter and 850 to 6,260 per square meter to 12.5 centimeters of mineral soil. Madge found 38,000 animals per square meter of litter during the wet season (of which 86 percent were mites and collembola) and only 400 during the dry season—a dramatic difference! Excluding mites and collembola, which are very small and will be considered separately, and multiplying the average estimate of numbers (7,500 per square meter) by a mean weight per individual of 0.25 milligrams dry weight (Edwards, 1967) resulted in an average biomass of 1.88 grams per square meter.

Table 2.8.     Live weight of birds collected by mist nets
in Panama forests.

| Species | Individual weight (grams) |
|---|---|
| Long-tailed hermit | 6.3 |
| White-flanked antwren | 7.7, 7.1 |
| Checker-throated antwren | 8.8, 8.5 |
| Sulphur-rumped flycatcher | 10.1 |
| Half-collared gnatwren | 11.7 |
| Olive-striped flycatcher | 12.7 |
| White-breasted woodwren | 14.9 |
| Spotted antbird | 16.5, 15.7, 13.1, 13.3, 14.9 |
| Slaty ant shrike | 20.5, 22.1 |
| Long-tailed woodcreeper | 21.8, 24.3 |
| Bicolored antbird | 27.3 |
| Scaly-throated leafscraper | 29.2 |
| Summer tanager | 29.8 |
| Swainson's thrush | 30.1 |
| Bare-crowned antbird | 31.9 |
| Gray-headed tanager | 33.0, 33.1, 30.7, 28.6 |
| Wing-banded antthrush | 37.6 |
| Wedgebilled woodcreeper | 39.8 |
| White-whiskered puff bird | 40.5 |
| Ocellated antbird | 43.0 |
| Crimson-crested woodpecker | 192.1 |
| Keeled-billed toucan | 343.0 |
| Average weight of an individual | 37.0 |

Microarthropods populations were estimated from data in Salt (1952), Williams (1941), Strickland (1947), De La Cruz (1964), Wiegert (1968). Their data suggest that populations range between 41,000 and 228,000 and average 112,000 individuals per square meter of soil surface to a depth of one meter. The mean number of microarthropods was multiplied by an estimated weight per individual of 0.03 milligrams dry weight (Edwards, 1967) to give a biomass of microarthropods of 3.36 grams.

The reader will appreciate that these data on animal biomass are hypothetical. Assuming that animals have finite limits of distribution in a habitat and that a sample more often than not will represent average rather than extreme densities, then estimates constructed from literature records such as these may be a reasonable basis for speculation. Interestingly Odum (1970), estimating animal biomass for the Montane forest at El Verde, Puerto Rico, found an animal biomass of 11.8 grams dry weight per square meter, an estimate that was quite similar to ours. Later we will examine the effect of different biomass of animals on the forest mineral cycling and will then be able to judge the sensitivity of the forest system to different densities of animals.

# CHAPTER 3

# The Chemical Content of the Forest

The chemical content of the forest is established in two steps. First, the quantity of an element in a given quantity of plant or animal material is determined. These concentrations are derived from the samples collected in the harvest study. Second, the concentrations are multiplied by the biomass to give the chemical inventory of a representative area of forest. Eight samples of plant material weighing about 500 grams were collected from each component, such as leaves, stems, and litter, in the harvest study, dried at 100°C. to constant weight, ground in a Wiley Mill and preserved in plastic bags for analysis. In the laboratory several methods were used to determine concentrations of the elements. The metallic elements were determined by atomic absorption spectrophotometry at the Savannah River Ecology Laboratory of the University of Georgia, Institute of Ecology, and by emission spectrometry at the Soil and Plant Analysis Laboratory of the University of Georgia Cooperative Extension Service. Nitrogen was determined by microkjeldhal techniques. Phosphorus was determined calorimetrically. In emission spectrophotometry dried samples were ashed at 500°C. and a solution of lithium carbonate and nitric acid was added to the ash before analysis. In this procedure the analysis was standardized several times daily and a blank was run after every 15 or 20 samples. Most vegetation samples were analysed by emission spectrophotometry.

In atomic absorption spectrophotometry a gram sample of dried plant and animal tissue was digested with 15 milliliters of a sulfuric-nitric-perchloric acid mixture, was diluted, filtered, and a lanthinum solution was added to the digested sample before submitting the sample to the spectrophotometer. Soil was extracted with the mixed acid solution under agitation for 30 minutes. All soil, animal, and water chemical analyses were determined by atomic absorption spectrophotometry which gave data on potassium, calcium, magnesium, sodium, cobalt, cesium, copper, iron, manganese, lead, strontium, and zinc. Also, cobalt, cesium, lead, and strontium in vegetation were determined by this method.

The spectrophotometry data sheets from the computer presented the data as percent of dry weight for the macroelements potassium, calcium, magnesium and sodium and as parts per million (ppm) dry weight for the micronutrients and trace elements. In the tables this order has been retained for convenience and all data converted to ppm.

### CONCENTRATIONS IN VEGETATION

The overall mean concentration of elements was similar in September and February vegetation samples (table 2.9). The Duncan's New Multiple Range Test of differences only showed statistical differences at the 0.5 level of probability for calcium, boron, barium,

Table 2.9.   Overall mean concentration of elements
in Tropical Moist forest vegetation in September and February.
(Concentration in ppm.)

| Element | September | February |
|---|---|---|
| N | 14000 | * |
| P | 1200 | 1200 |
| K | 11400 | 10700 |
| Ca | 16600 | 19000 |
| Mg | 1900 | 2000 |
| Na | 200 | 200 |
| Al | 1040 | 1059 |
| B | 17 | 27 |
| Ba | 96 | 37 |
| Co | 51 | 31 |
| Cs | 18 | * |
| Cu | 7 | 9 |
| Fe | 133 | 218 |
| Mn | 71 | 56 |
| Mo | 4 | 5 |
| Ni | 7 | * |
| Pb | 36 | * |
| Sr | 78 | 103 |
| Ti | 6 | 13 |
| Zn | 26 | 28 |
| Total, excluding N | 32890 | 34686 |

* Not measured.

Table 2.10.   Concentration of chemical elements (ppm)
The overall mean is based on all

| Compartment | P | K | Ca | Mg | Na | Al | B |
|---|---|---|---|---|---|---|---|
| Overstory leaves | 2200 ab | 14300 a | 21200 b | 2700 a | 200 ab | 1002 abc | 38 a |
| Understory leaves | 2400 a | 16200 a | 20600 b | 2900 a | 100 b | 1680 a | 21 b |
| Overstory stems | 500 e | 12100 a | 10800 cd | 1000 c | 200 ab | 712 bc | 8 c |
| Understory stems | 800 de | 12500 a | 11400 cd | 1600 bc | 100 b | 900 bc | 10 c |
| Overstory fruits and flowers | 2200 ab | 17400 a | 6300 d | 2200 ab | 100 b | 1218 abc | 12 bc |
| Litter | 1500 c | 2400 c | 26100 a | 1400 bc | 100 b | 1386 ab | 22 b |
| Roots | 1100 cd | 7600 b | 14800 c | 2200 ab | 300 a | 472 c | 8 c |
| Overall mean | 1200 | 11400 | 16600 | 1900 | 200 | 1040 | 17 |

*Columns sharing the same letter are not statistically different at the 0.05 level.

Table 2.11.   Concentration of chemical elements (ppm)
Cesium and lead were not determined on the material from this site.

| Compartment | P | K | Ca | Mg | Na | Al | B |
|---|---|---|---|---|---|---|---|
| Overstory leaves | 800 c | 11200 d | 22700 b | 2100 ab | 200 b | 482 c | 63 a |
| Understory leaves | 600 c | 19500 b | 27100 a | 2600 ab | 400 a | 1767 a | 35 b |
| Overstory stems | 300 d | 5400 e | 11200 cd | 1400 d | 100 c | 420 c | 8 d |
| Understory stems | 300 d | 8000 e | 13200 c | 1200 d | 100 c | 1331 ab | 9 d |
| Overstory fruits and flowers | 1800 a | 16100 c | 8100 d | 1800 bd | 100 c | 371 c | 20 c |
| Understory fruits and flowers | 2100 a | 24400 a | 7500 cd | 2600 ab | 100 c | 490 ab | 18 cd |
| Litter | 100 d | 5600 e | 28900 a | 2500 ab | 200 b | 1556 ab | 42 b |
| Roots | 100 d | 6900 e | 21300 b | 2600 a | 200 b | 1501 ab | 12 cd |
| Overall mean | 1200 | 10700 | 19000 | 2000 | 200 | 1059 | 27 |

*Columns sharing the same letter are not statistically different at the 0.05 level.
**Not included in the statistical test.

in Tropical Moist forest sampled during September.
data for seven compartments.*

| Ba | Co | Cs | Cu | Fe | Mn | Mo | Pb | Sr | Ti | Zn |
|---|---|---|---|---|---|---|---|---|---|---|
| $108_a$ | $56_b$ | $32_a$ | $6_b$ | $63_b$ | $76_{ab}$ | $4.3_b$ | $41_{ab}$ | $95_a$ | $6_{bc}$ | $19_a$ |
| $122_a$ | $101_a$ | $22_c$ | $5_b$ | $128_b$ | $63_{ab}$ | $1.8_b$ | $36_{dc}$ | $98_a$ | $7_b$ | $24_a$ |
| $61_b$ | $35_{cd}$ | $21_c$ | $5_b$ | $30_b$ | $113_a$ | $2.5_b$ | $21_c$ | $50_{bc}$ | $4_c$ | $25_a$ |
| $108_a$ | $35_{cd}$ | $27_b$ | $7_{ab}$ | $100_b$ | $47_{ab}$ | $1.1_b$ | $32_{bc}$ | $62_b$ | $5_c$ | $24_a$ |
| $39_b$ | $36_{cd}$ | $28_b$ | $12_a$ | $68_b$ | $64_{ab}$ | $16.4_a$ | $31_{bc}$ | $28_c$ | $4_c$ | $25_a$ |
| $137_a$ | $44_c$ | $22_c$ | $9_{ab}$ | $393_a$ | $103_a$ | $7.1_b$ | $48_c$ | $114_a$ | $9_a$ | $30_a$ |
| $69_b$ | $35_{cd}$ | $22_c$ | $11_a$ | $119_b$ | $25_b$ | $1.0_b$ | $40_{ab}$ | $71_b$ | $6_{bc}$ | $32_a$ |
| 96 | 51 | 18 | 7 | 133 | 71 | 4.0 | 36 | 78 | 6 | 26 |

in Tropical Moist forest sampled during February.
The overall mean is based on all data from the eight compartments.*

| Ba | Co | Cu | Fe | Mn | Mo | Sr | Ti | Zn |
|---|---|---|---|---|---|---|---|---|
| $28_b$ | $20_d$ | $7_{de}$ | $46_c$ | $23_b$ | 4.8 | $88_{cd}$ | $7_c$ | $21_a$ |
| $30_b$ | $26_c$ | $10_{cd}$ | $275_b$ | $49_b$ | 6.8 | $139_b$ | $19_b$ | $35_a$ |
| $19_b$ | $36_b$ | $4_f$ | $25_c$ | $56_b$ | 1.7 | $66_{de}$ | $5_c$ | $18_a$ |
| $18_b$ | $38_b$ | $5_{ef}$ | $62_c$ | $30_b$ | 1.8 | $65_{de}$ | $6_c$ | $16_a$ |
| $17_b$ | $36_b$ | $17_a$ | $97_c$ | $42_b$ | 6.6 | $48_e$ | $6_c$ | $45_a$ |
| $9_b$ | $49_a$ | $16_{ab}$ | $79_c$ | $60_b$ | 1.1 | $54_{de}$ | $4_c$ | $36_a$ |
| $108_a$ | $33_b$ | $12_{bc}$ | $513_a$ | $118_a$ | 5.9 | $188_a$ | $20_b$ | $28_a$ |
| $34_b$ | $34_b$ | $9_d$ | $493_a$ | $66_b$ | 4.1 | $117_{bc}$ | $26_a$ | $34_a$ |
| 37 | 31 | 9 | 218 | 56 | 3.6 | 103 | 13 | 28 |

cobalt, nickel, and strontium. Except for barium and cobalt, the forest sampled in September had lower overall mean concentrations, where the observed difference was statistically significant. Highest concentrations were determined for calcium, nitrogen, and potassium.[1]

Within forests, the greatest concentrations were in fruits, flowers and leaves (tables 2.10 and 2.11). Understory leaves and fruits often contained higher concentrations than overstory materials. Stems had lowest concentrations. For elements such as calcium, magnesium, aluminum, barium, iron, manganese, strontium, and titanium, roots or litter had relatively high concentrations.

### CONCENTRATIONS IN SOILS

The chemical concentrations in soils were determined solely by atomic absorbtion spectrophotometry and thus some trace element concentrations are not available for comparison with the vegetation. The soils of the two Tropical Moist forest sites differed more than did the vegetation (table 2.12), with Rio Sabana soil containing greater total concentrations of soil elements than that at Rio Lara. Maximum concentrations in both locations were observed for calcium and magnesium. Except for cobalt, copper, iron, and manganese all of the observed differences were statistically significant ($P > 0.5$).

We do not have adequate background information to explain these observed differences in the soils. They may represent seasonal effects, site differences, and/or reflect sampling error. However, a comparison of the soil data with that from several studies of soils of Tropical Moist forest in Panama (Gamble et al., 1969; Blue et al., 1969) and Costa Rica (Holdridge, 1964) shows that generally our two sets of values fall within the range found by others. Since our samples represent total element concentration rather than ammonium acetate (exchangeable) extractable concentration we would expect our values to be higher. The comparison shows that total concentrations were greater for iron and zinc at both locations, for sodium at Rio Sabana, and copper at Rio Lara (table 2.13). However, the total concentrations of strontium are considerably below the exchangeable quantities; the Rio Sabana potassium total concentration is also low. Generally the soil concentrations seem to be similar to those of the other investigations; however, strontium may be underestimated.

---

[1] The reader should note that these data differ from those reported in preliminary reports of Golley et al. (1969) and McGinnis et al. (1969) based on the untested chemical analyses.

Table 2.12. Average concentration (ppm) of chemical elements
in the soil of Tropical Moist forest. Samples represent total element
concentration to a depth of 30 centimeters.

| Element | Rio Lara | Rio Sabana |
|---------|----------|------------|
| P | 7.6 | 2.6 |
| K | 117 | 41 |
| Ca | 4232 | 5854 |
| Mg | 645 | 521 |
| Na | 50 | 462 |
| Co | 1.6 | 1.5 |
| Cs | 1.6 | * |
| Cu | 2 | 1 |
| Fe | 6 | 6 |
| Mn | 8 | 6 |
| Pb | 2.2 | * |
| Sr | 16 | 20 |
| Zn | 51 | 10 |
| Total | 5140.0 | 6925.1 |

* Not determined.

Table 2.13. Comparison of chemical concentration (ppm)
of Tropical Moist forest soils in Panama and Costa Rica.

| Exchangeable or total concentration | P. | K | Ca | Mg | Na | Cu | Fe | Mn | Sr | Zn | Authority |
|---|---|---|---|---|---|---|---|---|---|---|---|
| Total | 2.6 | 41 | 5854 | 521 | 462 | 1 | 6 | 6 | 20 | 10 | This study, Rio Sabana |
| Total | 7.6 | 117 | 4232 | 645 | 50 | 2 | 6 | 8 | 16 | 51 | This study, Rio Lara |
| Exchangeable | 9.1 | 650 | 2420 | 700 | —— | 1 | 2 | 15 | 214 | 2 | Gamble et al., 1969 |
| Exchangeable | 14.0 | 850 | 4700 | 1620 | —— | 1 | 1 | 22 | 72 | 3 | Gamble et al., 1969 |
| Exchangeable | 1.7 | 606 | 7880 | 1144 | —— | —— | —— | —— | 55 | —— | Blue et al., 1969 |
| Exchangeable | 0.7 | 80 | 7500 | 1330 | —— | 1 | 2 | 1 | 251 | 3 | Gamble et al., 1969 |
| Exchangeable | 1.1 | 82 | 1534 | 600 | 85 | —— | —— | 19 | —— | —— | Holdridge, 1964 |

We add a further comment about lead in the soil since there is current interest in this element. Kehoe (1961) described lead contents of several virgin soils in Mexico, Yucatan, and Sarawak as ranging from 0.07 to 30.0 mg/kg day sample. Urban soils in Cincinnati ranged from 28 to 360 mg per kg. According to Edwards (1971) the world average is 16 ppm. Clearly the undisturbed Darien, Panama, soils (2.2 ppm) fall within the estimate for virgin soil and are well below the world average.

ANIMAL CONCENTRATIONS

Animals were preserved in formalin or alcohol in the field and were later dried and prepared for chemical analysis. Vertebrates were disected and samples of flesh and bone were analyzed separately. Small organisms were analyzed as whole animals. Mammals were represented by aguoti (*Dasyprocta punctata*), birds by crested guan (*Penelope purpurascens*), reptiles by iguana (*Iguana sp.*), insects were sweep net samples of vegetation-dwelling insects, detritivores were represented by termites.

The chemical concentrations for animals (table 2.14) showed some important differences between animal parts and taxa. Calcium, magnesium, colbalt, and sodium levels were high in bone. Iron was high in iguana flesh. Insects contained relatively large quantities of zinc and sodium.

These data on the individual animals were converted to average concentrations for animal taxonomic groupings and then to trophic levels. To do this, it was assumed that 21 percent of the dry weight of mammal was bone (Kaufman et al., 1963), that bone was 67 percent of the dry weight of birds (Noles, personal communication) and reptiles. Insect data could be used directly; termites were used to represent detritivores. The average values are shown in table 2.15. For almost every element there are relatively large differences among trophic levels reflecting differences between the taxa used in the analysis. We suspect that there is considerable chemical diversity in the fauna of the forest, although this supposition contradicts the findings of Beyers et al. (1970), who suggest that animals are integrators of chemical diversity in the vegetation.

Table 2.14.  Elemental concentration in animals from tropical forests in Panama (ppm dry weight).

| Taxon | P | K | Ca | Mg | Na | Co | Cs | Cu | Fe | Mn | Pb | Sr | Zn |
|---|---|---|---|---|---|---|---|---|---|---|---|---|---|
| Agouti, soft tissue | 846 | 4503 | 2146 | 1561 | 3162 | 354 | 285 | 273 | 4064 | 51 | 253 | 28 | 246 |
| Agouti, bone | 417 | 1379 | 47379 | 5069 | 7105 | 1371 | 203 | 86 | 2507 | 75 | 513 | 103 | 271 |
| Crested Guan, soft tissue | 2075 | 8035 | 4826 | 2330 | 51C7 | 501 | 381 | 242 | 4297 | 153 | 231 | 46 | 527 |
| Crested Guan, bone | 638 | 3317 | 27701 | 5460 | 9265 | 1713 | 307 | 333 | 3854 | 175 | 539 | 119 | 815 |
| Iguana, soft tissue | 2494 | 3091 | 240 | 1277 | 1084 | 565 | 240 | 203 | 7809 | 36 | 391 | 17 | 325 |
| Iguana, bone | 67 | 217 | 18206 | 1005 | 1358 | 1739 | 173 | 67 | 1521 | 47 | 271 | 81 | 114 |
| Termites | 487 | 2697 | 7880 | 2590 | 4886 | 63 | 212 | 35 | 1607 | 165 | 114 | 50 | 133 |
| Insects on ground vegetation | 395 | 5527 | 6450 | 3522 | 9545 | 13 | — | 124 | 1461 | 103 | 12 | 14 | 758 |

Table 2.15.    Average concentrations (ppm dry weight) for animal groups in a tropical forest in eastern Panama.

| Animal group | P | K | Ca | Mg | Na | Co | Cu | Fe | Mn | Pb | Sr | Zn |
|---|---|---|---|---|---|---|---|---|---|---|---|---|
| Mammals | 756 | 3847 | 11645 | 2298 | 3990 | 568 | 234 | 3737 | 56 | 308 | 44 | 251 |
| Birds | 1112 | 4874 | 16111 | 4427 | 7893 | 1313 | 303 | 4000 | 168 | 437 | 95 | 720 |
| Reptiles and amphibians | 870 | 1165 | 12313 | 1095 | 1268 | 1352 | 112 | 3596 | 43 | 311 | 60 | 184 |
| Insects | 395 | 5527 | 6450 | 3522 | 9545 | 13 | 124 | 1461 | 103 | 12 | 14 | 758 |
| Detritivores | 487 | 2697 | 7880 | 2590 | 4886 | 63 | 35 | 1607 | 165 | 114 | 50 | 133 |
| Herbivores[1] | 648 | 4351 | 10087 | 2665 | 5657 | 402 | 201 | 3054 | 70 | 219 | 35 | 403 |
| Carnivores[2] | 817 | 1728 | 11697 | 1425 | 2182 | 1175 | 126 | 3387 | 51 | 289 | 56 | 239 |
| Detritivores[3] | 487 | 2697 | 7880 | 2590 | 4886 | 63 | 35 | 1607 | 165 | 114 | 50 | 133 |

[1]The herbivore biomass is calculated from the percentage representation of the groups in table 2.6
[2]The carnivore biomass is calculated from the percentage representation of the groups in table 2.6.
[3]Detritivores were represented by termites.

CONCENTRATION FACTORS

Concentrations in the forest ecosystem components can be compared as ratios of a given element in two compartments to show accumulation or exclusion. Petrographic analyses of Sabana Shales (IOCS-FD-60, 1968) provide information on several elements in the parent material. The ratios of potassium, calcium, magnesium and sodium in the surface soil to that in the Sabana shale (table 2.16) show that at both sites there is an enrichment of these cations in the soil. The exception to this conclusion is sodium at the Rio Lara site. The sodium concentration at this site is 50 ppm or about a factor of ten less than the concentration at Rio Sabana which is near the range of sodium in soils from Bowen (1966). Probably the Rio Lara soil sodium levels are low; if so, the ratios of soil to rock and vegetation to soil will be similar at both sites.

The ratios of concentrations of potassium, calcium, magnesium and sodium in vegetation to that in the soil show an enrichment of the vegetation for potassium, calcium, and magnesium and an exclusion of sodium (table 2.16). The enrichment is much greater for potassium than for calcium and magnesium. The concentration factors for the two sites are similar.

The ratios of concentration in herbivores to vegetation (table 2.16) show a discrimination for potassium and calcium, and an enrichment for magnesium and sodium. Sodium is concentrated in animals much more than is magnesium. All other element concentrations in vegetation to soils showed enrichment. Enrichment ratios were greatest for phosphorus followed by iron and cobalt. Comparison of concentrations in herbivores to vegetation showed a discrimination for phosphorus and strontium and an enrichment for all other elements tested.

These concentration ratios reflect essentiality as well as abundance of the elements in the environment. Clearly the surface soil, which is a dynamic part of the ecosystem, contains a larger total concentration of the essential elements than the parent material, even with leaching from rainwater. Probably these higher concentrations are a function of the mineral cycle and the presence of organisms in the soil which act against the loss or export of essential elements. In this forest the surface soil acts as a nutrient reservoir to some extent.

The concentrations in vegetation and herbivores reflect the requirements for nutrients of the different kinds of organisms. The plants have high requirements for potassium and phosphorus and these elements are highly concentrated in the plant tissues. Calcium and magnesium are less concentrated since they appear in high concentrations in the soil. Animals, in contrast, require sodium which is discriminated against by the plants. Sodium is highly concentrated as a consequence by the herbivores.

### THE INVENTORY OF ELEMENTS

With the information on biomass and chemical concentrations in the biomass, we can calculate the total chemical inventory of the forests. The amount of an element in a compartment was calculated by

Table 2.16.  Concentration factors for four
elements in Tropical Moist forest.

| Ratio | Element | | | |
|---|---|---|---|---|
| | K | Ca | Mg | Na |
| Rio Sabana | | | | |
| Herbivores/vegetation | 0.4 | 0.5 | 1.3 | 28.3 |
| Vegetation/soil | | 3.2 | 3.8 | 0.4 |
| Soil/parent material | 2.3 | 27.7 | 5.9 | 3.0 |
| Rio Lara | | | | |
| Herbivores/vegetation | 0.4 | 0.6 | 1.4 | 28.3 |
| Vegetation/soil | 97 | 3.9 | 2.9 | 4.0 |
| Soil/parent material | 6.5 | 20.1 | 7.2 | 0.3 |

multiplying the compartment biomass by its elemental concentration. The weight of the soil compartment to a depth of 30 centimeters was determined by using the conventional estimate of bulk density of 1.47 grams per cubic centimeter. This average is very similar to that (1.56) reported by Greenland and Kowal (1960) for the top 30 centimeters of forest soil in Africa and for forest soil by Singh (1967) in India and Tsutsumi et al. (1966) in Thailand, but is much higher than that of Popenoe (1959) for the upper 5 to 10 centimeters of forest soil in Guatemala (0.56 g/cc) and Jordan (1971) for El Verde, Puerto Rico, Montane forest soil. An overestimate of bulk density would overestimate the quantities of elements in the soil. The values presented here differ from those reported in Golley et al. (1969) and represent the concentrations determined by emission spectrophotometry. The present values should be considered as correct and final.

Table 2.17.   Comparison of the inventory of chemical elements (kg/hectare) in Tropical Moist forest soil and vegetation on two sites.

| Element | Rio Lara | | Rio Sabana | |
|---|---|---|---|---|
| | Vegetation | Soil | Vegetation | Soil |
| P | 241 | 33 | 85 | 11 |
| K | 4598 | 508 | 1606 | 197 |
| Ca | 4702 | 18582 | 3502 | 25749 |
| Mg | 437 | 2830 | 423 | 2281 |
| Na | 78 | 220 | 31 | 2023 |
| Al | 296 | —— | 144 | —— |
| B | 4 | —— | 3 | —— |
| Br | 2b | —— | 6 | —— |
| Co | 14 | 7 | 10 | 7 |
| Cs | 8 | 7 | —— | —— |
| Cu | 2 | 9 | 1 | 4 |
| Fe | 20 | 26 | 16 | 26 |
| Mn | 43 | 35 | 16 | 26 |
| Mo | 1 | —— | T | —— |
| Sr | 22 | 70 | 20 | 87 |
| Ti | 2 | —— | 2 | —— |
| Zn | 10 | 225 | 5 | 42 |
| Total | 10504 | 22552 | 5870 | 30453 |

Comparison of the inventory in the vegetation and the soil at the two sites suggests that there is a large site or seasonal effect on the chemical content of the forest (table 2.17). Wet season vegetation

Table 2.18.  Percentage of the total
chemical inventory stored in the vegetation
in two Tropical Moist forests.

| Element | Wet season, Rio Lara site | Dry season, Rio Sabana site |
|---|---|---|
| P | 88 | 89 |
| K | 90 | 89 |
| Ca | 20 | 12 |
| Mg | 13 | 16 |
| Na | 26 | 2 |
| Co | 67 | 59 |
| Cs | 53 | — |
| Cu | 18 | 20 |
| Fe | 43 | 38 |
| Mn | 55 | 38 |
| Sr | 24 | 19 |
| Zn | 4 | 12 |
| All vegetation combined | 31 | 16 |

contained larger quantities of total elements and larger absolute quantities for all elements except titanium. In contrast, dry season soils contained larger total quantities of elements; this difference was due to the dry season site having larger quantities of calcium and sodium. These differences are in the expected direction. The effects of leaching during the wet season and leaf fall during the dry season result in lower inventories in dry season vegetation and higher inventories in dry season soils. However, the differences observed in table 2.17 are not only a function of season but also of differences in biomass and in elemental concentration between sites. The biomass-plot effect is probably more important than concentration effect since the quantities of biomass are large with respect to the concentrations. The wet season Rio Lara plot contained the largest biomass (377,807 kg/ha compared to 276,131 kg/ha) and this difference was not entirely a seasonal effect since the amount of woody material (stems

Table 2.19.   Inventory of elements in
Quantities in kilograms per hectare are the average

| Component | P | K | Ca | Mg | Na | Al | B | Ba |
|---|---|---|---|---|---|---|---|---|
| Leaves | 16 | 135 | 221 | 25 | 2 | 9 | 0.5 | 0.7 |
| Stems | 128 | 2846 | 3355 | 357 | 48 | 182 | 2.4 | 13.2 |
| Fruits and flowers | 0.2 | 1.3 | 0.5 | 0.2 | t* | 0.1 | t | t |
| Roots | 6 | 81 | 208 | 27 | 3 | 12 | 0.1 | 0.5 |
| Litter | 14 | 39 | 318 | 20 | 2 | 17 | 0.3 | 1.5 |
| Soil to 30 cm. | 22 | 353 | 22166 | 2256 | 1121 | —— | —— | —— |
| Herbivores | .01 | .05 | .13 | .03 | .07 | —— | —— | —— |
| Carnivores | .01 | .01 | .10 | .01 | .02 | —— | —— | —— |
| Detritivores | .03 | .14 | .41 | .14 | .26 | —— | —— | —— |
| Total | 186.3 | 3455.5 | 26269.1 | 2685.4 | 1176.4 | 220.1 | 3.3 | 15.9 |

*$t$ means less than one-tenth of a kilogram in vegetation or soil per hectare, or one-hundredth of a kilogram of animal.

and roots), which would not change with season, on the Rio Sabana
site was 73 percent of the Rio Lara woody biomass. For comparison
leaves and fruits on the Rio Sabana plot were 67 percent of the
Rio Lara leaf and fruit biomass. Apparently, rather more than one-
half of the difference in the chemical inventory of the vegetation between seasons is due to differences in woody biomass. The differences
in soil inventories are entirely due to differences in concentration since
the same bulk density was used for soils on both sides.

The inventory in vegetation and soil at the two sites can be combined
to determine the percentage of the total inventory in the vegetation
compartments (table 2.18). This comparison will provide an answer
to one of our original questions—does the tropical forest concentrate
a large proportion of the chemical inventory in the living biomass as
a nutrient conserving adaptation? Considering all elements the answer
must be no, since the vegetation contains only 16 to 30 percent of the
total chemical inventory in the Tropical Moist forest. However,
phosphorus and potassium do not follow the overall pattern, having
more than 80 percent of their inventory in the vegetation. Cobalt,
cesium, and manganese also appear at levels over 50 percent in the
vegetation. The overall pattern is mainly due to the calcium, mag-

components of the Tropical Moist forest.
of the quantities in the wet and the dry season.

| Co | Cs | Cu | Fe | Mn | Mo | Pb | Sr | Ti | Zn |
|------|------|-----|------|------|-----|------|------|-----|-------|
| 0.5 | 0.4 | 0.1 | 0.7 | 0.5 | 0.1 | 0.5 | 1.0 | 0.1 | 0.2 |
| 10.8 | 7.4 | 1.4 | 8.6 | 27.2 | 0.6 | 7.4 | 17.4 | 1.3 | 6.7 |
| *t* | *t* | *t* | *t* | *t* | *t ,* | *t* | *t* | *t* | *t* |
| 0.2 | 0.2 | 0.1 | 3.7 | 0.5 | *t* | 0:4 | 1.1 | 0.2 | 0.4 |
| 0.5 | 0.4 | 0.1 | 5.0 | 1.2 | 0.1 | 0.8 | 1.5 | 0.1 | 0.4 |
| 7.0 | 7.0 | 6.5 | 26.0 | 30.5 | —— | 10.0 | 78.5 | —— | 133.5 |
| .01 | —— | *t* | .04 | *t* | —— | *t* | *t* | —— | .01 |
| .01 | —— | *t* | .03 | *t* | —— | *t* | *t* | —— | *t* |
| *t* | —— | *t* | .08 | .01 | —— | .01 | *t* | —— | .01 |
| 19.0 | 15.4 | 8.2 | 44.2 | 59.9 | 0.8 | 19.1 | 99.5 | 1.7 | 141.2 |

nesium, and sodium levels in the soil and may be partly due to our use of total rather than exchangeable soil element content for the comparison. This is not the entire explanation of the pattern, however, since the exchangeable concentration will be less than the total concentration but probably not sufficiently less to change the pattern, (see table 2.13). Apparently the proportion of the chemical inventory held in the tropical forest biomass varies with the type of soil, as well as with intensity of precipitation and other factors.

For later calculations it will be necessary to have a value of the chemical content of each component in the forest, excluding the seasonal or site differences. We have averaged the two inventories and combined them with the animal inventories for the total forest chemical content (table 2.19). The animal inventory was determined by multiplying the concentrations of trophic groups (table 2.15) by their biomass (table 2.6). In the total inventory calcium is most abundant and most of the calcium occurs in the soil compartment. Potassium is next most abundant, with most potassium in stems. Magnesium is almost as abundant as potassium and is mainly in soil. Animals contain an extremely small amount of each element compared to vegetation or soils.

# III
# DYNAMICS OF THE TROPICAL MOIST FOREST

*The biological cycling of ash elements and nitrogen is one of the major aspects of the interrelationships between plants and soil, the components of a biogeocenosis that most clearly define its limits and express its nature.*

L. E. RODIN AND N. J. BAZILEVICH

In the previous section, the Tropical Moist forest has been described in static terms as consisting of quantities of chemical materials per area of land surface. These quantities specify the amounts stored in each component of the system at an instant of time. To complete the description of cycling in the forest, we also must know the rate at which materials are transferred from component to component.

For this study we assumed that chemicals move through the forest system in three main ways: in the fall of organic material from the canopy to the ground, through leaching by rain water passing through the leaves and flowing over the branches and stems, and by translocation from the soil to the roots, stems and leaves. It is possible to intercept the organic matter or litter and the water falling from the canopy and, therefore, to measure the quantities of chemicals moving under the force of gravity. If we assume that the major movement of chemicals is in a cyclic pattern of roots-stem-leaves-litter-soil-roots and that the forest is not growing and storing materials in the biomass, then the amount moving by gravity flow must equal the amounts being translocated up the stem. The movement of chemicals to and through animal populations also can be calculated separately from knowledge of the amounts of food consumed by the trophic groups. Each of these separate transfers can be identified on the system block diagram, which is illustrated in fig. 1.3.

These assumptions are not strictly true, of course. There are transfers in the reverse direction from leaves to stems and roots, but in the absence of data on these biological reverse transfers in tropical trees we have made the simplifying assumption of unidirectional flow. The forest steady state assumption has also been questioned but the implications of this assumption can and will be tested with the model later in this section.

Besides the transfers within the forest it is also necessary to account for the inputs and outputs to the system. Chemical materials enter the forest by two main routes, rainfall and weathering of the parent material. The major output is to streams draining the forest watershed. In this section, the transfers into, within and out of the Tropical Moist forest will be presented sequentially. First, data on the cycle within the forest will be discussed. Then, data on system inputs and outputs will be described. Finally, the system cycle and the inputs-outputs will be combined into a generalized forest model, which will be subjected to several analyses.

CHAPTER 4

# Forest Litter

Litter includes leaves, stems, branches, fruits, flowers, and other plant parts, as well as some animal remains and fecal material. Once the litter falls, it undergoes a process of breakdown or decay with the eventual release of the chemical materials comprising the organic tissues. In humid tropical forests, leaves and floral parts may begin the decay process while still on the plant, and decomposition, with the exception of certain highly resistant woods, proceeds rapidly on the soil surface. The rate at which litter falls and decays can be determined by collecting the falling organic material in suitable traps and by isolating the fallen material on the ground and reexamining it at intervals.

## PROCEDURES

Litter was collected on ten one-meter plots during the harvest of each site. About one year later, a second sample of ten plots was taken at the same location to determine if the quantity of litter on the ground changed from one year to the next. This measurement tested the assumption that in a mature forest the litter biomass would be in equilibrium and would not increase or decrease in a year.

Ten litter boxes, constructed of wood and plastic screening, were placed in each mature forest type (fig. 3.1) to collect falling organic material. The collection screen was about 20 centimeters above the ground. Litter, generally made up of leaves, small twigs, fruits, and flowers, was collected from the boxes at about two-month intervals and was weighed and dried. Samples of the litter were ground in a Wiley mill and sent to the University of Georgia for chemical analysis.

Established in each forest were ten ground plots, two by two meters in size, from which branch material greater than 2 centimeters diameter was collected and weighed. Samples were taken for determination of dry weight and for chemical analysis. Actually the litter boxes also collected some larger branches but this material has not been treated separately.

In the Tropical Moist forest, ten one-meter square plots were estab-

Figure 3.1.   A litter collection box in the forest. The box is approximately a meter square and is held 20 cm above the forest floor by wooden legs. (Photograph by authors.)

lished to determine litter decomposition rates. In February, fresh leaves were placed on clean plots, but these packed and formed a putrefying mass that was totally unlike natural litter. Therefore, in May new plots with natural litter were established. Litter was collected and weighed as it came from the ground surface. About 300 g (dry weight) were put on a square meter. The plot was covered with a plastic screen, held about 10 cm above the litter, to prevent addition of fresh litter to the plots (fig. 3.2). Dry weights were determined from other samples which were taken to the laboratory for oven drying. At two-month intervals, plots were selected for harvest; the litter was removed, bagged, and returned to the laboratory for drying and weighing. The progressive change in weight of the litter plots established in May, 1967, gave a measure of the breakdown of organic matter.

Figure 3.2.   A litter decomposition plot in the Tropical Moist forest.
A fiberglass screen was held above the plot to prevent addition of
falling litter. Plots were one-meter square in size. (Photograph by
authors.)

### LITTER FALL AND DECOMPOSITION

The standing crop of litter in grams dry weight per square meter,
reported earlier, was 620 during the dry season and 291 during the
wet season. Approximately one year later, the Tropical Moist forest
during the dry season had 736 grams per square meter on the forest
floor. There was no significant difference in the standing crop of litter
during the same season between the two years by the t-test, supporting
our expectation that the litter in these forests is in equilibrium. Comparisons made in the Premontane Wet forest which is located on mountainous terrain in Darien Province, Panama, gave similar results; litter
on ten plots in February was 482 and 551 grams per square meter in
1967 and 1968. The dry season Tropical Moist forest standing crop esti-

mates also were similar to those reported by Woods and Gallegos (1970) for Tropical Moist forest at another location in Panama.

Table 3.1.   Litter and branch fall in Tropical Moist forest
(kg dry weight per hectare).

| Date collected | Number of days between collections | branches | Leaf litter | Total | Rate per day |
|---|---|---|---|---|---|
| May 2 | 89 | 701 | 3120 | 3821 | 43 |
| June 15 | 44 | 35 | 1068 | 1103 | 25 |
| August 24 | 70 | 132 | 2158 | 2290 | 33 |
| October 29 | 66 | 41 | 1039 | 1080 | 16 |
| January 12 | 75 | 445 | 1682 | 2127 | 28 |
| February 27 | 46 | 294 | 1434 | 1728 | 38 |
| Total | 390 | 1647 | 10500 | 12147 | |

Litter and branch fall measured in the litter boxes and on the twig plots varied with season (table 3.1). Leaf litter was almost ten times branch litter over the year and the greatest total daily rate of litter fall occurred during the dry season. Lowest inputs of litter occurred in the August-October period. This pattern is similar to that observed by Cunningham (1963), Nye (1961), and Madge (1965) in African forests.

A comparison of the yearly litter fall to the litter standing crop gives information on litter turnover. Considering both leaf and branch litter, yearly dry matter production in kilograms per hectare was 11,350 in the Tropical Moist forest. When the standing crop of litter during the dry season (6,200 kg/ha 1967 and 7,360 in 1968) is compared with litter fall, the ratio of litter fall to standing crop is 1.5 and 1.8 in the two years. However, when this ratio is calculated using the standing crop of litter in the wet season (2,910 kg/ha), it is 3.9, or almost double the dry season ratio. Possibly this is the reason for the wide variation in the ratio of litter fall to litter standing crop reported for other tropical forests (table 3.2). Regardless of the base used in comparison, the data clearly show that litter turnover is usually less than one year.

Another way to examine these relationships is to compare the litter fall of leaves as measured by the litter boxes and the standing crop of

Table 3.2.   Comparison of annual litter fall in tropical forests
(kg dry wt/ha/yr).

| Forest | Litter fall | Standing crop litter | Ratio of litter fall to standing crop | Authority |
|---|---|---|---|---|
| Tropical Moist forest | 11350 | 6200* | 1.83 | This study |
| Premontane Wet forest | 10480 | 4820 | 2.17 | This study |
| Riverine forest | 11610 | 14150 | 0.82 | This study |
| Colombian Rainforest | 8520 | 5040 | 1.69 | Jenney et al. (1949) |
| Colombian Rainforest | 10110 | 16480 | 0.61 | Jenney et al. (1949) |
| Panamanian Second-Growth | 6000 | 8650 | 0.69 | TTC (1966) |
| Ghana Semideciduous | 10536 | 2264 | 4.65 | Nye (1961) |
| Nigerian Dry Lowland forest | 5600 | 1700-2450 | 2.29-3.29 | Madge (1965) |
| Nigerian Moist Semideciduous forest | 4625 | 1715 | 2.70 | Hopkins (1966) |
| Nigerian Evergreen forest | 7170 | 3040 | 2.36 | Hopkins (1966) |

*Dry season.

leaves on trees. The biomass of leaves on the trees was greater than the fall of leaves during the wet season, but was less in the dry season, because the deciduous species lose their leaves during the dry season. The ratio of litter fall to biomass of leaves was 1.24 during the dry season and 0.82 during the wet season, suggesting that leaf turnover on the trees is probably slightly greater than one year. In a similar comparison in Thailand rain forest, Kira et al., (1967) found the ratio of litter fall to leaf standing crop to be 1.45.

Litter decomposition on plots established in May, 1967, showed that eight to nine months after establishment only 9 percent of the original material in the Tropical Moist forest and 17 percent in the Premontane Wet forest remained on the plot. These data support the conclusion derived from the comparison of standing crop and fall of litter that litter turnover is less than one year.

Breakdown of litter on the decomposition plots was not linear (fig. 3.3). While the records from the Tropical Moist forest varied somewhat erratically from month to month, presumably because of the movement of dirt and debris into the litter by animals, the general trend was a slow rate of decay for the first six months followed by a period of rapid decay, after October (fig. 3.3). Witkamp and Olson (1963), studying de-

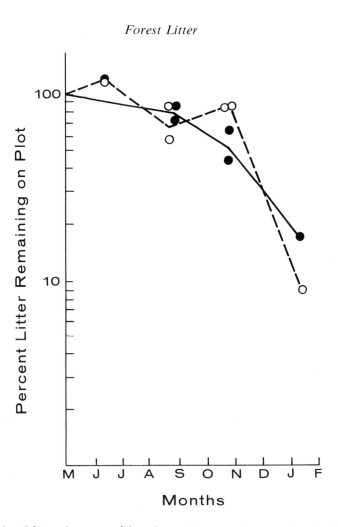

Figure 3.3.  Litter decomposition from May to January in the Tropical Moist (broken line) and Premontane Wet (solid line) forests.
The percentage of the original weight is plotted against time. Each
point represents one 1m² plot.

composition of litter in Tennessee using leaves strung on string (which
simulates the normal position of leaves on the ground), also found an
initial period of slow decay followed by a period with a higher decay
rate. In contrast, Madge (1965) in Nigeria found a rapid and relatively
constant rate of decay of leaf discs confined in meshbags during the
wet season. While leaf decay almost stopped during the dry season of
four months, the leaf tissue had entirely disappeared in a total of seven

months. Wiegert and Murphy (1970) report significant effects of season and species of leaf on litter decomposition in Puerto Rico. Hopkins (1966) also states that decomposition of tropical litter takes place in an average of five months (range 2.5-7.2 months). Obviously, the decay process is under control of temperature, moisture, and leaf composition and will vary between forests within the tropics, but all these studies agree that litter turnover in tropical forests is less than one year. In contrast, temperature forest litter turnover rates may be 1/4 to 1/64 that of the tropical forests (Olson, 1963).

STANDING CROP OF CHEMICAL ELEMENTS IN THE LITTER

We would expect that the concentration of chemicals in the litter would be less than that in the leaves since the litter has experienced decomposition and leaching. In the Tropical Moist forest, phosphorus and potassium were the only elements conforming to this pattern (table 3.3); for all other elements litter contained similar or greater concentrations than leaves. Differences between leaves and litter were significant for phosphorus, potassium, calcium, strontium, iron, and manganese. Higher litter concentrations may be due to contamination of the litter by soil through the activity of organisms, rain splash, or the collecting procedure. In the Premontane Wet forest, phosphorus, potassium, sodium, and cobalt concentrations in litter were significantly less than leaf concentrations, while calcium, strontium, and iron were significantly higher in litter.

Another independent measure of litter chemical concentrations was samples from the litter boxes. Interestingly these samples deviated even further from the expected pattern (table 3.3), with concentrations of phosphorus, magnesium, sodium, and cobalt in Tropical Moist forest and phosphorus in the Premontane Wet forest greater than either leaves or litter. In other words, in these instances litter in the boxes gained chemical content or was composed of something other than leaves. We know that all falling material is caught by the boxes, that fruits and flowers make up the greatest proportion of the non-leaf material, and that fruits and flowers often have a higher chemical content than leaves. Nevertheless, when the elemental input to the box is recalculated on the assumption that the fruits and flowers and leaves are falling in proportion to their standing crops, the values, while higher, are still not very similar to those in the litter boxes. Something else is contributed to the boxes, possibly the material is insect debris—we do not know.

Table 3.3. Comparison of the elemental concentration in overstory leaves, litter from the litter boxes and litter on the ground. Data in ppm, based on grams dry weight.

| Site | Compartment | P | K | Ca | Mg | Na | Co | Cs | Cu | Fe | Mn | Pb | Sr | Zn |
|---|---|---|---|---|---|---|---|---|---|---|---|---|---|---|
| Tropical Moist Rio Sabana | Overstory leaves | 800 | 11200 | 22700 | 2100 | 200 | 20 | — | 7 | 46 | 23 | — | 88 | 21 |
| | Litter box | 1556 | 2574 | 17543 | 3085 | 699 | 40 | 12 | 6 | 132 | 91 | 25 | 162 | 27 |
| | Litter | 100 | 5600 | 28900 | 2500 | 200 | 33 | — | 12 | 513 | 118 | — | 188 | 28 |
| Premontane Wet | Overstory leaves | 294 | 10200 | 11000 | 3716 | 2800 | 80 | 39 | 7 | 259 | 371 | 48 | | |
| | Litter box | 804 | 1204 | 10250 | 2325 | 1077 | 30 | 15 | 5 | 130 | 244 | 33 | 152 | 48 |
| | Litter | 160 | 5000 | 14700 | 2300 | 300 | 39 | 46 | 13 | 1079 | 369 | 36 | 173 | 39 |

Concentration of the chemical elements in branches from the Tropical Moist forest for phosphorus, potassium and calcium were greater than those in Premontane Wet forest (table 3.4).

CHEMICAL TRANSFER IN LITTER

The movement of elements from the canopy to the forest floor can be calculated from the concentration of overstory leaves, the major contributor to litter, and the quantities falling per year. The branch contribution can be determined from the branch concentrations and the branch fall per year. The annual input to the litter is the sum of these two inputs. The gain or loss from the litter to the soil can be calculated as the difference between the litter input and the litter standing crop.

The annual inputs to the litter are greatest for calcium, potassium, and magnesium (table 3.5). Greater quantities of phosphorus, potassium, and calcium occur in litter input of Tropical Moist forest, while for the other elements annual input is greater for the Premontane Wet forest. About 250 kilograms of these 13 elements are added to the forest litter annually.

Table 3.4.   Concentration (ppm) and the amounts of selected elements in annual branch fall (kg/ha) in Tropical Moist (Rio Lara site) and Premontane Wet forest.

| Elements | Tropical Moist forest overstory stems | | Premontane Wet forest overstory stems | |
|---|---|---|---|---|
| | Concentration | kg/ha | Concentration | kg/ha |
| P | 500 | 0.77 | 80 | 0.11 |
| K | 12100 | 18.63 | 5800 | 7.68 |
| Ca | 10800 | 16.63 | 5900 | 7.81 |
| Mg | 1000 | 1.54 | 1100 | 1.46 |
| Na | 200 | 0.31 | 1400 | 1.85 |
| Co | 35 | 0.05 | 49 | 0.06 |
| Cs | 21 | 0.03 | 27 | 0.04 |
| Cu | 5 | 0.01 | 5 | 0.01 |
| Fe | 30 | 0.05 | 71 | 0.09 |
| Mn | 113 | 0.17 | 308 | 0.41 |
| Pb | 21 | 0.03 | 30 | 0.04 |
| Sr | 50 | 0.08 | 62 | 0.08 |
| Zn | 25 | 0.04 | 40 | 0.05 |

Table 3.5. Dynamics of litter (kg/ha) in two Panamanian tropical forests.

| | P | K | Ca | Mg | Na | Co | Cs | Cu | Fe | Mn | Pb | Sr | Zn |
|---|---|---|---|---|---|---|---|---|---|---|---|---|---|
| **Tropical Moist** | | | | | | | | | | | | | |
| Annual input to the litter | 8.6 | 128.7 | 239.7 | 22.2 | 2.3 | 0.3 | — | 0.1 | 0.5 | 0.4 | — | 0.9 | 0.3 |
| Standing crop of litter | 14.0 | 39.0 | 318.0 | 20.0 | 2.0 | 0.5 | — | 0.1 | 3.7 | 1.2 | — | 1.5 | 0.4 |
| Difference between input and standing crop | +5.4 | -89.7 | +78.3 | +2.2 | -0.3 | +0.2 | — | 0.0 | +3.2 | +0.8 | — | +0.6 | +0.1 |
| **Premontare Wet** | | | | | | | | | | | | | |
| Annual input to the litter | 2.6 | 90.6 | 97.7 | 32.9 | 24.9 | 0.7 | 0.3 | 0.1 | 2.3 | 3.3 | 0.04 | 1.4 | 0.4 |
| Standing crop of litter | 0.8 | 24.1 | 70.9 | 11.1 | 1.5 | 0.2 | 0.2 | 0.1 | 5.2 | 1.8 | 0.2 | 0.8 | 0.2 |
| Difference between input and standing crop | -1.8 | -66.5 | -26.8 | -21.8 | -23.4 | -0.5 | -0.1 | 0.0 | +2.9 | -1.5 | -0.2 | -0.6 | -0.2 |

Table 3.6.    Comparison of annual chemical input to forest floor in leaf litter fall
in different forests.

| Vegetation | Release in kg/ha/yr. | | | | Authority |
|---|---|---|---|---|---|
| | Calcium | Magnesium | Potassium | Phosphorus | |
| Panama | | | | | |
| Tropical Moist | 240 | 22 | 129 | 9 | This study |
| Premontane Wet | 98 | 33 | 91 | 3 | This study |
| Riverine | 165 | 29 | 110 | 13 | This study |
| Mangrove | 87 | 34 | 60 | 6 | This study |
| India | | | | | |
| *Terminalia—Shorea* | 184 | 30 | 26 | 28 | Singh (1968) |
| *Tectona* | 120 | 11 | 20 | 8 | Singh (1968) |
| *Tectona* plantation | 131 | 5 | 19 | 11 | Seth et al. (1963) |
| *Diospyros—Anogeissus* | 84 | 17 | 31 | 3 | Singh (1968) |
| *Shorea robusta* plantation | 77 | 10 | 19 | 9 | Seth et al. (1963) |
| *Shorea—Buchanania* | 32 | 11 | 8 | 3 | Singh (1968) |
| Ghana | | | | | |
| 40-year-old forest | 206 | 45 | 68 | 7 | Nye (1961) |
| Temperate Forests | | | | | |
| Beech Gap, Tennessee | 37 | 9 | 33 | 5 | Shanks et al. (1961) |
| Spruce fir, Tennessee | 19 | 6 | 14 | 4 | Shanks et al. (1961) |
| Spruce, Russia | 46 | 6 | 17 | 2 | Cited by Ovington (1965) |
| Scots pine, England | 49 | 9 | 57 | 10 | Ovington (1959) |

Chemical input from leaves was significantly greater than the litter standing crop, except for calcium, strontium, iron, and manganese in Tropical Moist forest and iron in Premontane Wet forest. In these cases, the elements occur in relatively high concentrations in the soil which suggests there has been movement of soil into the litter, or contamination of litter with soil. This is the same conclusion we came to in the examination of concentrations in leaves and litter (table 3.3). The effect of leaching of the litter is apparent especially for phosphorus, potassium, and sodium. Also, the Premontane Wet forest showed a more widespread pattern of leaching loss from litter than the Tropical Moist forest.

Movement of chemical elements through branch fall can be added to those quantities moving in the litter. Branch fall amounts to 1,541 kg/ha/yr in Tropical Moist forest and 1,324 kg/ha/yr in Premontane Wet forest. The quantities of chemicals moving in this pathway are much lower than those transferred in leaf litter (table 3.4).

To put the data on movement of minerals through litter fall into perspective, the data from Panama has been compared to that from other forests in table 3.6. The annual release of calcium, magnesium, and phosphorus compares with that reported for tropical forests in India and Africa, while the release of potassium is higher than for other tropical or temperate forests. Release of calcium, magnesium, and potassium in Panamanian tropical forests also is higher than that in temperate forests. These differences mainly reflect the differences in leaf biomass in temperate and tropical forests. Leaf biomass is greater in tropical forests, and therefore, the quantities of chemicals reaching the litter annually are also greater.

# CHAPTER 5

# Throughfall

Water enters the Tropical Moist forest as rain. Some rain striking the irregular surface of the canopy is held by the leaves, stems, branches, and attendant epiphytes and evaporates; the remainder flows through the leaves and stems to the forest floor. Chemical materials may be carried by the rain into the forest, but the rain also leaches chemicals from the leaves and stems. This throughfall provides an alternative source of chemical elements to the litter.

Rain was measured with standard rain gages. Measurements to 0.025 millimeters were recorded daily at the Santa Fe base camp between 19 June and 31 December 1967. Monthly records were collected from February to June at Santa Fe. These and additional rainfall records from the National Oceanographic and Atmospheric Association (NOAA) and the Panama Canal Company hydrology section were used to compute the average rainfall for the Sabana River watershed in the Tropical Moist forest near Santa Fe.

Studies of the quantities of rainfall intercepted by the canopy were conducted during the week of 18 September 1967. The study was located between the Santa Fe base camp and the Lara River and was adjacent to the Rio Lara harvest site. Five plots were selected, located between 5 and 10 meters apart. Based on ten or more separate estimates, the plots had average canopy coverage of 42, 57, 93, and 98 percent. A cleared area was selected as an input reference. At each of the six locations, a throughfall gage was constructed from a 1.2 by 1.8 meter plastic sheet suspended over the forest floor, and sloped to the center at one end. Interception areas for each suspended sheet were calculated and the collection sensitivities were computed to be 0.0025 mm. The average collection area was 1.986 $m^2$. The three rains occurring during the course of the study occurred on 18 September, 0.20 mm; 19 September, 33.94 mm; and 20 September, 1.72 mm. Samples of the water collected under the canopy and in the open were analyzed for chemical content.

The lightest rain barely penetrated the canopy (table 3.7), while the heaviest rain was so intense we could not complete all the measurements inside the forest. The data from the 0.20 and 1.12 mm rains (table 3.7) suggest that percent interception is directly related to percent cover of the canopy; the 58 percent cover plot is the only one which deviated from this trend. While a direct relation between cover and rain interception may be true for light rains, the percentage of heavy rains intercepted by the vegetation is dependent only on the initial quantity required to saturate the canopy. When this quantity is intercepted all rains run off the leaves and branches.

In the Tropical Moist forest, the average canopy cover was about 81 percent and most daily rains (70 percent) were less than 6.2 mm (table 3.8). Therefore, it is reasonable to assume that for the general case, a rain of about 3.8 mm (the midpoint of the 1.3-6.2 mm class, table 3.8) was required to saturate the canopy and, for these rains, 81 percent was intercepted. We further assume that rainfall amounts in excess of 8.8 millimeters passed through the canopy as throughfall with little or no interception. These assumptions ignore the very complicated relationship between rain intensity, throughfall, and cover of the vegetation, but in the absence of more exact measurements they are probably reasonable approximations. They also ignore the flow of water down stems; however, Nye (1961) reported that less than one percent of the rain flowed down trunks of moist tropical forest in Ghana. We also noted little stem flow, even during heavy rains in eastern Panama forests.

Using these assumptions and considering the rainfall reported at Santa Fe (table 3.9) as input to the forest, 19 percent (367 mm) of

Table. 3.7.   Interception of rain by Tropical Moist forest canopy.

| Canopy cover percent | Rain of 0.20 millimeter | | Rain of 1.12 millimeter | |
|---|---|---|---|---|
| | Percent interception | Deviation from % canopy cover | Percent interception | Deviation from % canopy cover |
| 0 | 0 | 0 | 0 | 0 |
| 42 | 49 | + 7 | 39 | - 3 |
| 58 | 93 | +35 | 41 | -17 |
| 92 | 99 | + 6 | 97 | + 4 |
| 95 | 99 | + 4 | 98 | + 3 |
| 98 | 99 | + 1 | 98 | 0 |

Table 3.8.   Frequency and percentage of daily rainfall
for 170 daily observations at the Santa Fe base camp.

| Daily rainfall mm | Frequency days |
|---|---|
| 0.0-1.2 | 81 |
| 1.3-6.2 | 38 |
| 6.3-12.5 | 20 |
| 12.6-18.7 | 11 |
| 18.8-25.0 | 5 |
| 25.1-31.2 | 6 |
| 31.3-37.5 | 2 |
| 37.6-43.7 | 4 |
| 43.8-50.0 | 1 |
| Over 50.1 | 2 |

Total 3.9.   Total monthly rainfall at Santa Fe,
in the Tropical Moist forest.

| Month | Millimeters/month |
|---|---|
| January* | 20 |
| February* | 32 |
| March* | 0 |
| April* | 51 |
| May* | 102 |
| June | 393 |
| July | 426 |
| August | 130 |
| September | 242 |
| October | 208 |
| November | 222 |
| December | 107 |
| Total | 1933 |

* Data obtained from hydrology section, **Panama Canal Company.**

Table 3.10.  Chemical composition in ppm of rain throughfall
collected at ground level in the Tropical Moist forest.

| Observation number | P | K | Ca | Mg | Na | Co | Cu | Fe | Mn | Pb | Sr | Zn |
|---|---|---|---|---|---|---|---|---|---|---|---|---|
| 1 | 0.036 | 3.0 | 2.0 | 0.44 | 1.56 | 0.06 | 0.010 | 0.400 | 0.018 | 0.024 | 0.004 | 0.032 |
| 2 | 0.024 | 2.4 | 2.8 | 0.72 | 1.52 | 0.06 | 0.018 | 0.216 | 0.010 | 0.016 | 0.004 | 0.030 |
| 3 | 0.032 | 5.6 | 2.4 | 0.56 | 1.52 | 0.06 | 0.010 | 0.300 | 0.020 | 0.024 | 0.006 | 0.046 |
| 4 | 0.036 | 2.4 | 2.4 | 0.60 | 1.36 | 0.06 | 0.016 | 0.416 | 0.036 | 0.016 | 0.004 | 0.029 |
| 5 | 0.068 | 2.6 | 2.4 | 0.80 | 1.72 | 0.07 | 0.030 | 0.036 | 0.026 | 0.016 | 0.004 | 0.062 |
| Average | 0.039 | 3.2 | 2.4 | 0.62 | 1.54 | 0.06 | 0.017 | 0.273 | 0.022 | 0.019 | 0.004 | 0.040 |

the input rainfall (1,933 mm) would be intercepted and evaporated by
the Tropical Moist forest. This estimate is similar to that of Odum
(1967) for Lower Montane forest in Puerto Rico (29 percent) and is
only slightly higher than the 8 to 16 percent reported for temperate
forests (Boggess, 1956; Semago and Nash, 1962; Helvey and Patric,
1965; and Lawson, 1967), which have one-half the leaf surface area
and a lower amount of rainfall. It is less, however, than the estimates
of 43 percent interception by other tropical forests in Puerto Rico
(Clegg, 1963), and 66 percent for an evergreen forest in Brazil (Freise,
1936).

Table 3.11.  Transfer of elements from the vegetation to the soil
in the Tropical Moist forest (kg/ha/yr).
The vegetation inventory is based on table 2.19.

| | P | K | Ca | Mg | Na | Co | Cu | Fe | Mn | Sr | Zn |
|---|---|---|---|---|---|---|---|---|---|---|---|
| Litterfall | 8.6 | 129 | 240 | 22 | 2.3 | 0.3 | 0.05 | 0.5 | 0.4 | 0.9 | 0.3 |
| Branch fall | 0.8 | 19 | 17 | 2 | 0.3 | 0.05 | 0.03 | 0.1 | 0.2 | 0.1 | 0.04 |
| Rainwash | 0.6 | 50 | 37 | 10 | 24.0 | 1.0 | 0.26 | 4.3 | 0.3 | 0.06 | 0.6 |
| Total | 10.0 | 198 | 294 | 34 | 26.6 | 1.4 | 0.34 | 4.9 | 0.9 | 1.06 | 0.94 |
| % of inventory in above-ground vegetation | 6.9 | 6.6 | 8.2 | 8.9 | 53.3 | 14.9 | 22.6 | 52.8 | 3.2 | 5.8 | 13.6 |

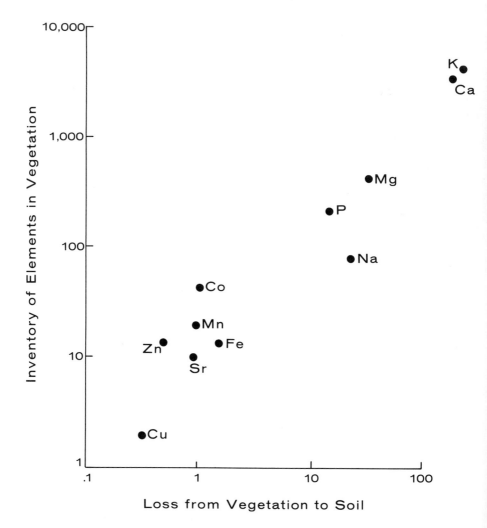

Figure 3.4.   Relationship between the inventory of chemical elements (kg/ha) and the annual transfer from vegetation to soil (kg/ha/yr).

The chemical concentration of the throughfall was dominated by potassium, calcium, and sodium (table 3.10). On the basis of studies of Tukey and Tukey (1962), we would expect leaching from the leaves in the order Na>Mg>K >Sr >Fe>Zn>P. Attiwell (1966 also showed that sodium and potassium are readily leached, but states that phosphorus, magnesium, and calcium are held in the leaves. In our samples, calcium seems to deviate from this pattern but as we will show later calcium also appears in large quantities in the rainwater.

The throughfall quantity was 81 percent of the annual rainfall or $15.6 \times 10^6$ kg/ha/yr. This quantity was multiplied by the concentrations in throughfall to give the quantities of elements moving by this route. A comparison of transfer of chemicals by throughfall and litter and branch fall (table 3.11) shows that litter fall is the most important route of transfer except for sodium, iron, cobalt, copper, and zinc which moved in greatest amounts in throughfall. These latter elements also exhibited the greatest percentage transfer out of the vegetation inventory (table 3.11). It seems very unlikely that over 50 percent of the sodium and iron inventory would turn over each year. Without comparative data, we cannot explain these results, but possibly the analysis of the throughfall water for these two elements was in error. In absolute terms, the greatest transfer from the vegetation to the soil is for calcium, potassium, magnesium, and sodium.

Since this transfer is from the vegetation, the rate of transfer should be related to the quantity of elements stored in the vegetation (fig. 3.4). Clearly there is a linear relationship between transfer rate and the inventory; the average turnover of all elements is 13 percent per year.

# CHAPTER 6

# Transfer to Animal Populations

In this study, it was not possible to quantify the transfers from litter to soil, to roots, and from roots to stems to leaves. For purposes of modelling, we assumed that the forest is in steady state and is not gaining or losing chemical content. Comparative data on forest biomass (section 2) and litter fall support this assumption. If steady state conditions exist, then the uptake of roots and stems (fig. 1.3) must approximately equal throughfall plus litter fall since the leaves, litter, soil, roots, stems, leaves form a linear sequence of transfers. The transfers to animals form complications on this sequence (fig. 1.3) and must be evaluated separately.

Chemical materials enter the animal populations through consumption of food. Since food intake is influenced by the kind of animal and its body size, it was possible to estimate input rates for each taxonomic and size category. The biomass for each taxa within food habit groups has been presented in table 2.6. The average dry weight of individual animals was calculated from our samples or from the literature as described in section 2, and is listed in table 3.12. The food intake of each size category representative of each animal group was determined from the literature. Intake for an average mammal was derived from figure 9-2 in Davis and Golley (1963). Bird intake was calculated from data of Nice (1938). Reptile and amphibian intake was estimated from feeding experiments by Duever (1967), supported by data of Johnson (1966) and McNab (1963). Insect food intake was estimated from Reichle (1968), and microarthropods from Englemann (1961). These feeding rates are summarized in table 3.12.

Next, intake to each trophic group (table 3.13) was calculated by multiplying the population biomass (table 2.6) by the feeding rate (table 3.12) and summing intake for all taxa within the group. Herbivores and detritivores have the largest intakes, frugivores and carnivores the least. Carnivore consumption is about a factor of 10 lower than herbivore consumption. Herbivores and frugivores together consumed about 0.3 g/m²/day, or, over a year, about 10 percent of the

Table 3.12.   Body weight of representatives of animal groups
and food intake (weight food/body weight/day).

| Animal group | Average weight per individual (dry weight) | | Food intake | |
|---|---|---|---|---|
| Mammals | 1158.0 | g | 0.33 | g |
| Birds | 11.0 | g | 0.14 | g |
| Reptiles and amphibians | 1.5 | g | 0.011 | g |
| Vegetation arthropods | 2.5 | mg | 0.054 | mg |
| Ground macroarthropods | 0.25 | mg | 0.063 | mg |
| Ground microarthropods | 0.03 | mg | 0.025 | mg |

Table 3.13.   Intake of food ($g/m^2/day$) by the animal
populations in the Tropical Moist forest.

| | Food intake | |
|---|---|---|
| **Herbivores** | | |
| Mammals | 10.234 | |
| Arthropods | 10.023 | |
| Total | | 0.257 |
| **Frugivores** | | |
| Mammals | 0.026 | |
| Birds | 0.001 | |
| Arthropods | 0.001 | |
| Total | | 0.028 |
| **Carnivores** | | |
| Mammals | 0.013 | |
| Birds | 0.003 | |
| Reptiles and amphibians | 0.008 | |
| Arthropods | 0.003 | |
| Total | | 0.027 |
| **Detritivores** | | |
| Soil and litter macroarthropods | 0.118 | |
| Soil and litter microarthropods | 0.084 | |
| Total | | 0.202 |

Table 3.14.   Elemental intake of animals (kg/ha/yr).

| Animal group | | P | K | Ca | Mg | Na | Al | B | Ba | Co | Cs | Cu | Fe | Mn | Mo | Pb | Sr | Ti | Zn |
|---|---|---|---|---|---|---|---|---|---|---|---|---|---|---|---|---|---|---|---|
| | | | | | | | | | | | | Elements | | | | | | | |
| Herbivores | | | | | | | | | | | | | | | | | | | |
| | Consuming leaves | 2.06 | 13.41 | 19.89 | 2.53 | 0.19 | 0.94 | 0.04 | 0.10 | 0.05 | 0.03 | 0.01 | 0.06 | 0.07 | 0.004 | 0.04 | 0.09 | 0.006 | 0.02 |
| | Consuming fruit | 0.22 | 1.77 | 0.64 | 0.22 | 0.01 | 0.12 | 0.00 | 0.00 | 0.00 | 0.00 | 0.00 | 0.01 | 0.01 | 0.002 | 0.00 | 0.00 | 0.001 | 0.003 |
| Carnivores | | 0.06 | 0.43 | 0.99 | 0.26 | 0.56 | — | — | — | 0.04 | — | 0.02 | 0.30 | 0.01 | — | 0.02 | 0.00 | — | 0.04 |

standing crop of leaves and fruit. Intake by the detritivore group, 0.202 $g/m^2/day$ was underestimated since this complex of organisms must consume the litter and dead animals in the system. The estimated feeding rate for arthropod detritivores (about 700 kg/ha/year) would not be sufficient to consume the litter fall of about 11,000 kg/ha/yr. Based on estimates of $O_2$ metabolism by Zeuthen (1953), we calculate a population bimass of one gram per square meter of bacteria or about $10^{12}$ bacteria would be required to consume the standing crop of litter in one year. Coulter (1950) found populations of 28,000,000 bacteria per gram of soil in tropical forests in Malaya. Converting his estimate to a square meter basis, using a bulk density of soil of 1.47, gives an estimate of $2 \times 10^{12}$ bacteria to a depth of 5 cm. Populations of this size would be sufficient to consume the litter annually available in Panamanian forests. Since our estimates of detritivores obviously underestimate the transfers from litter to soil, we have assumed linear flow from litter to soil through the component.

The intakes of herbivores and carnivores were calculated by multiplying the animal biomass intake per day (table 3.13) by 365 days and then by the concentration of elements in ppm in overstory leaves and fruit in the Rio Lara sample (table 2.10) or by the herbivore concentrations (table 2.15). The intakes for each group are presented in table 3.14.

The ratio of herbivore and frugivore to carnivore consumption varies widely for the elements. For example, phosphorus, potassium, calcium, magnesium, manganese, and strontium intake in carnivores is 10 percent or less than that of herbivores. In contrast, for the other elements carnivore intake is 50 (lead) to 500 (iron) percent of herbivore consumption. These differences are due to the different concentrations of the elements in plant and animal tissues.

# CHAPTER 7
# System Inputs and Outputs

The input of chemical materials carried in the rain may be an important contribution to the forest system. Rain measured by a standard rain gage, at Santa Fe base camp was used as the input to the Tropical Moist forest. Other gages operated by NOAA and the Panama Canal Company in the Darien Province give comparative data for the Santa Fe record. Rain input to the Sabana River watershed of 259 km² was 1933 mm/m² (table 3.15) during the year, with greatest rainfall occurring in June and July.

The frequency of daily rainstorm intensities (table 3.8) was determined between 19 June and 31 December 1967. One hundred seventy separate daily observations were recorded at the Santa Fe base camp rain gage. Forty-eight percent (81) of the days had 1.2 mm or less of rain, of which 15 days were rain free and 34 days had only a trace amount of 0.025 mm. Ninety-one percent of the rains were less than 25.0 mm, about eight percent between 25.0 and 50.0 mm and one percent greater than 50.0 mm. The maximum daily rainfall during the period was 76.71 mm (3.02 inches).

The seasonal distribution of rainfall is not uniform across eastern Panama (fig. 3.5). During 1967, rain fell almost continuously over approximately nine months from mid-April to late December, preceded by three months of less or no rainfall. At certain locations rainfall decreased in September and October, as described generally for Central America by Portig (1965); this two-period form of rainfall intensity is illustrated most clearly in the mountainous regions at Cuadi, Nurra, and Esloganti (fig. 3.5). Tropical Moist forest occurs at stations Morti to Sabana on the figure. The general trend for the total annual rainfall across the isthmus is an increase from about 2,000 mm along the Pacific coast to 3,600 mm in the Caribbean coastal mountains, and then a decrease to about 2,800 mm along the Caribbean coast.

The chemical composition of rain water falling during the canopy interception study in September, 1967, is reported in table 3.16. The

Table 3.15.  Continuous daily rainfall record (mm per day) from
Santa Fe base camp during 1967.

| Days of month | June | July | Aug. | Sept. | Oct. | Nov. | Dec. |
|---|---|---|---|---|---|---|---|
| 1 | | 0.0 | 0.0 | 1.8 | 6.1 | 2.5 | 0.3 |
| 2 | | 0.0 | 7.6 | 0.0 | 2.5 | 30.2 | 1.3 |
| 3 | | 19.6 | 0.0 | 26.2 | *t* | 7.4 | 1.8 |
| 4 | | 1.3 | 2.5 | *t* | *t* | 3.0 | 27.9 |
| 5 | | 15.7 | 18.3 | *t* | *t* | 0.3 | 10.9 |
| 6 | | 0.5 | 1.5 | 9.1 | *t* | 11.2 | 3.6 |
| 7 | | 0.0 | 0.3 | 13.7 | 2.3 | 41.4 | *t* |
| 8 | | 40.4 | 0.0 | 1.5 | 19.8 | 3.0 | *t* |
| 9 | | 11.4 | | 46.7 | 31.0 | 1.0 | 1.3 |
| 10 | | 15.0 | | 5.6 | *t* | 11.7 | *t* |
| 11 | | 17.8 | | 0.5 | *t* | *t* | 7.6 |
| 12 | | 40.9 | | 0.5 | 14.2 | 1.0 | 16.0 |
| 13 | | 17 0 | | 10.2 | 15.2 | 7.1 | 24.1 |
| 14 | | 66.0 | 48.3* | 0.8 | | *t* | *t* |
| 15 | 326.6* | 5.6 | 0.3 | 2.3 | | 3.8 | *t* |
| 16 | | 1.3 | 0.3 | 4.8 | | 2.0 | *t* |
| 17 | | 2.0 | 1.3 | *t* | 41.9* | 16.8 | 2.3 |
| 18 | | 0.0 | 0.3 | *t* | *t* | 2.0 | *t* |
| 19 | 12.9* | 0.0 | 7.9 | *t* | *t* | 2.8 | *t* |
| 20 | 27.9 | 10.2 | 2.5 | 34.5 | | 3.8 | *t* |
| 21 | 0.8 | 76.7 | 0.3 | *t* | | 8.9 | 7.1 |
| 22 | 4.3 | 0.5 | 1.3 | 2.8 | | 19.1 | *t* |
| 23 | 3.3 | 31.8 | 2.0 | *t* | | 1.3 | 1.0 |
| 24 | 3.3 | 5.3 | 0.8 | | | 2.5 | 2.0 |
| 25 | 0.0 | 0.0 | 7.6 | | | 1.3 | 0.3 |
| 26 | 1.3 | 0.0 | 13.2 | | 30.0* | 12.7 | *t* |
| 27 | 6.4 | 2.5 | 2.8 | | 42.2 | 20.3 | *t* |
| 28 | 6.4 | 29.2 | 0.5 | 71.1* | 0.3 | 3.3 | *t* |
| 29 | 0.0 | 1.0 | 0.0 | 3.6 | | *t* | *t* |
| 30 | 0.0 | 2.0 | 1.3 | 5.8 | 0.8* | 1.3 | *t* |
| 31 | | 12.7 | 9.7 | | 1.5 | | *t* |

\* Accumulated total.
*t* Indicates trace amounts.

average values for calcium, potassium, sodium, phosphorus, and magnesium are within the range of the values reported by Nye (1961)
in Ghana, Robertson and Davies (1965) in United Kingdom, Emanuelsson, et al. (1954) in Sweden, and Junge and Werby (1958) in the

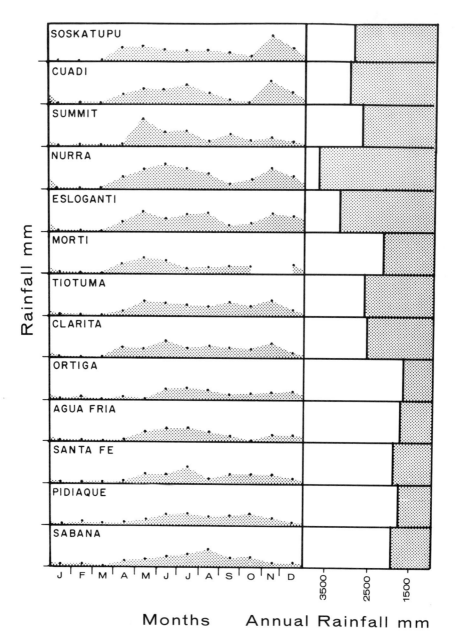

Figure 3.5. Monthly rainfall in 1967 from 13 stations on a transect across eastern Panama. Stations on the Caribbean coast are at the top of the figure; those on the Pacific side of the isthmus are at the bottom.

United States, but are above the average for continental United States. Mineral content in precipitation depends upon distance from the sea, uptake from soil in dust, dilution in the atmosphere, and other factors. We might expect above-average levels of potassium, calcium, and sodium in rainfall on an isthmus between two oceans.

If we compare these rain concentrations with the throughfall (table 3.10), we see that potassium, magnesium, iron, and calcium were greater in throughfall than rain water—that is, these elements leached from the vegetation. The other elements occurred at similar or lower levels in throughfall than in rainwater. For comparison, Nye (1961) showed that throughfall in a Ghana rain forest was a factor of 17 greater for potassium, 10 for phosphorus, 3 for calcium, and 2 for magnesium over the quantities in rainfall.

Table 3.16. Chemical composition of rainfall in ppm in eastern Panama for September, 1967.

| Observation number | P | K | Ca | Mg | Na | Co | Cu | Fe | Mn | Pb | Sr | Zn |
|---|---|---|---|---|---|---|---|---|---|---|---|---|
| 1 | 0.045 | 1.4 | 3.6 | 0.64 | 1.93 | 0.04 | 0.026 | 0.213 | 0.010 | 0.064 | 0.008 | 0.079 |
| 2 | 0.075 | 0.7 | 1.4 | 0.14 | 1.42 | 0.16 | 0.068 | 0.123 | 0.068 | 0.054 | 0.003 | 0.052 |
| 3 | 0.047 | 0.2 | 0.7 | 0.10 | 1.70 | 0.10 | 0.030 | 0.053 | 0.003 | 0.013 | 0.001 | 0.069 |
| 4 | 0.042 | 0.3 | 0.4 | 0.04 | 1.12 | 0.01 | 0.009 | 0.036 | 0.002 | 0.008 | 0.002 | 0.017 |
| 5 | 0.033 | 0.3 | 1.3 | 0.13 | 1.80 | 0.14 | 0.010 | 0.053 | 0.006 | 0.026 | 0.001 | 0.036 |
| 6 | 0.043 | 0.3 | 0.7 | 0.20 | 1.70 | 0.10 | 0.003 | 0.340 | 0.036 | 0.026 | 0.003 | 0.021 |
| 7 | 0.064 | 0.2 | 2.5 | 0.50 | 1.46 | 0.18 | 0.025 | 0.271 | 0.025 | 0.025 | 0.008 | 0.039 |
| Average | 0.050 | 0.5 | 1.5 | 0.25 | 1.59 | 0.10 | 0.024 | 0.155 | 0.021 | 0.031 | 0.004 | 0.044 |

SYSTEM OUTPUT IN STREAM WATER

Discharge from the Sabana River watershed was calculated from data in the IOCS Hydrology report (IOCS-FD-59, 1968) which reported 168,094 acre-feet discharge for the 259 square kilometers watershed area from May to December, 1967. The watershed above the gage is about 24 kilometers long and an average of 11.2 kilometers wide. The river slope varies from a fall of about 26 meters per kilometer, in the first two miles from its origin on the drainage divide, to about 3.8 meters per kilometer in the next 7.5 kilometers, then flattens to a fall of about 1.3 meters per kilometer. Acre-feet were converted to acre-inches and then to millimeters discharged (table 3.17) by dividing discharge by the acres on the watershed and converting to millimeters.

Table 3.17.   Relationship of monthly precipitation and stream discharge on the Sabana River Watershed, 1967.
Annual total was calculated on the assumption that at the Sabana gage the total from May to December was the same proportion of the annual flow as at Tiotuma gage on the Rio Chucunaque with a drainage area of 130 square miles, where the discharge during May to December was 92 percent of the total annual discharge.

| Months | Precipitation millimeter | Discharge millimeter | Percent |
|---|---|---|---|
| May | 102 | 7 | 7.0 |
| June | 393 | 71 | 18.0 |
| July | 426 | 112 | 26.3 |
| August | 130 | 70 | 53.8 |
| September | 242 | 58 | 24.0 |
| October | 208 | 201 | 96.6 |
| November | 222 | 209 | 94.1 |
| December | 107 | 61 | 57.0 |
| Total | 1830 | 789 | 43.1 |
| Annual | 1933 | 857 | 44.3 |

Rain input is also reported in table 3.17 for comparison with discharge. Input and discharge were not directly related; greatest rainfall was in June and July, and greatest discharge in October and November. This would be expected since a substantial portion of the rains falling at the beginning of the wet season should be stored in the soil. Input appearing in discharge increases from May to August and is usually less than 50 percent. In contrast, in October and November when the soil apparently has been recharged, input and discharge are almost equal.

Over the eight-month period, discharge was 43 percent of precipitation (table 3.17). When the Sabana data were extrapolated to an annual basis, using the full-year data from Tiotuma gage on the Rio Chucunaque, the discharge was 44 percent of precipitation. This estimate is supported by data of Matthew (1950) for the Gatun Lake watershed, Canal Zone, where discharge is 40 percent of annual precipitation. It is also similar to the 51 to 58 percent of Likens et al. (1967) for Hubbard Brook, New Hampshire, the 52 percent of Dils (1957) for Coweeta, North Carolina, and the 51 and 63 percent of Iwatsubo and Tsutsumi (1968) for broadleaf forest in Japan, but is considerably greater than the estimate of 29 percent by Odum (1967) for the Montane rain forest in El Verde, Puerto Rico, and of 17 to 20 percent by Pereira (1967) and Pereira et al. (1962) for East African rain forest.

River water was collected for chemical analyses at hydrology monitoring stations during the wet and dry seasons. The samples give data on the dissolved and suspended particulate material discharged from the watershed, but not the bed load. While we do not account for all the losses from the watershed, we probably do account for those associated with the forest and surface soils. Leopold et al. (1964) showed that the percent of the total load that is dissolved in streams is high where precipitation is great, runoff high, and vegetation abundant. For example, they reported 64 percent of the total load in solution for Juniata River, Pennsylvania, and Borman et al. (1969) reported 85 percent of Hubbard Brook, New Hampshire. Also Birot (1968) states that tropical rivers typically have a small quantity of rock fragment in the river bed.

The chemical composition of river water from streams draining the region of study varied relatively widely from river to river (table 3.18). However, there appeared to be reasonably close correspondence between samples from the same river, even when these were taken in the wet and dry seasons (see Chucunaque, for example, table 3.18). This suggests that the variation in water quality in table 3.18 is a function of location rather than season. Iwatsubo and Tsutsumi (1968) also showed in Japan that concentration of chemicals in river water did not change with seasonal stream discharge.

Templeton et al. (1969) also measured the chemical concentration of water for the same set of rivers in the same year (table 3.19). The concentration of macroelements, with the exception of phosphorus, were similar in both studies, but our estimates of copper, iron, manganese, lead, and zinc were considerably above those of Templeton et al. (1969) and also the world average, according to Bowen (1966). Possibly these differences are due to differences in the methods of analysis or technique. For example, some of our colleagues have suggested that sodium may have leached from the glassware used for flash evaporation of the water.

Table 3.18. Chemical composition in ppm of river water from eastern Panama. (N.S. indicates that the element was not determined.)

| Sources | P | K | Ca | Mg | Na | Co | Cu | Fe | Mn | Pb | Sr | Zn |
|---|---|---|---|---|---|---|---|---|---|---|---|---|
| Chucunaque River dry season | N.S. | 1.0 | 13.5 | 4.03 | 9.14 | N.S. | 0.003 | 0.061 | 0.005 | N.S. | 0.061 | 0.017 |
| Chucunaque River wet season | N.S. | 1.1 | 17.3 | 2.99 | 6.81 | N.S. | 0.005 | 0.094 | 0.007 | N.S. | 0.061 | 0.010 |
| Cuadi River dry season | 0.032 | 0.0 | 2.2 | 5.71 | 4.45 | 0.081 | 0.008 | 1.010 | 0.008 | 0.021 | 0.008 | 0.019 |
| Cuadi River wet season | 0.112 | 0.0 | 5.3 | 2.65 | 14.70 | 0.079 | 0.053 | 1.033 | 0.530 | 0.026 | 0.009 | 0.072 |
| Morti River wet season | 0.046 | 0.3 | 6.0 | 2.27 | 5.93 | 0.060 | 0.006 | 0.113 | 0.006 | 0.026 | 0.016 | 0.030 |
| Morti River dry season | 0.035 | 0.6 | 6.3 | 2.90 | 10.70 | 0.113 | 0.018 | 0.289 | 0.063 | 0.025 | 0.018 | 0.057 |
| Cuadi River tributary wet season | 0.080 | 0.5 | 70.0 | 12.00 | 16.80 | 0.060 | 0.040 | 0.140 | 0.005 | 0.040 | 0.010 | 0.060 |
| Sasardi on Cuadi River wet season | 0.248 | 2.1 | 8.5 | 5.04 | 7.21 | 0.045 | 0.038 | 2.093 | 0.038 | 0.062 | 0.013 | 0.124 |
| Chucunaque River Ortiga wet season | 0.060 | 1.6 | 18.3 | 6.50 | 11.00 | 0.100 | 0.091 | 6.166 | 0.133 | 0.066 | 0.033 | 0.146 |
| Chucunaque River Tiotuma wet season | 0.044 | 0.6 | 11.0 | 1.61 | 8.33 | 0.040 | 0.034 | 1.105 | 0.020 | 0.053 | 0.034 | 0.121 |
| Paradise on Morti River wet season | 0.041 | 0.3 | 4.4 | 5.21 | 10.66 | 0.076 | 0.006 | 1.650 | 0.006 | 0.025 | 0.009 | 0.048 |
| Sabana River | 0.152 | 4.5 | 65.5 | 10.02 | 23.74 | 0.128 | 0.134 | 0.322 | 0.134 | 0.326 | 0.313 | 0.066 |
| Mean all rivers | 0.085 | 1.1 | 19.0 | 5.08 | 10.79 | 0.078 | 0.036 | 1.173 | 0.080 | 0.067 | 0.049 | 0.064 |

Comparison of the concentration of elements in river water and precipitation showed that in almost every case, the concentration in rain water was materially lower than that in average river water. Gorham (1961) suggests that tropical rivers are generally poor in dissolved salts and he presents data from British Guiana showing that river water does not deviate greatly from precipitation. This generalization apparently is less true for eastern Panama.

Summarizing the water chemistry, in the Tropical Moist forest region the rain water contains relatively high concentrations of potassium, calcium, sodium, and other elements. After passing through the canopy the water was enriched with potassium, magnesium, iron, and calcium. The water leaving the system as stream discharge contained relatively large concentrations of potassium, calcium, magnesium, iron, and sodium. In every case but for cobalt, river water contained higher concentrations of the elements than did rain water.

Output from the system is calculated as the product of volume (liters) of water discharged times the average concentration in the

Table 3.19. Comparison of the chemical concentration (ppm) of river water in Darien, Panama, by two separate studies, with the average composition for the world rivers.

| Location | Panama | Panama | World-wide |
|---|---|---|---|
| Authority | This study | Templeton et al. (1969) | Bowen (1966) |
| P | 0.085 | 0.011 | 0.005 |
| K | 1.07 | 0.86 | 2.3 |
| Ca | 19.04 | 15.14 | 15.0 |
| Mg | 5.08 | 5.13 | 4.1 |
| Na | 10.79 | 7.36 | 6.3 |
| Co | 0.08 | —— | 0.0009 |
| Cu | 0.04 | 0.006 | 0.01 |
| Fe | 1.17 | 0.103 | 0.67 |
| Mn | 0.08 | 0.017 | 0.012 |
| Pb | 0.07 | —— | 0.005 |
| Sr | 0.05 | 0.055 | 0.08 |
| Zn | 0.05 | 0.014 | 0.01 |

Table 3.20.    System dynamics of the Tropical Moist forest.

| Element | Input kg/ha/yr | Discharge kg/ha/yr | Net change kg/ha/yr | Total inventory kg/ha | Turnover (yrs) |
|---|---|---|---|---|---|
| P  | 1.0  | 0.7   | + 0.3   | 186   | 266 |
| K  | 9.5  | 9.3   | + 0.2   | 3456  | 372 |
| Ca | 29.3 | 163.2 | -133.9  | 26269 | 161 |
| Mg | 4.9  | 43.6  | - 38.7  | 2685  | 62  |
| Na | 30.7 | 92.5  | - 61.8  | 1176  | 13  |
| Co | 2.0  | 0.7   | + 1.3   | 19    | 27  |
| Cu | 0.5  | 0.4   | + 0.1   | 8     | 20  |
| Fe | 3.0  | 10.1  | - 7.1   | 44    | 4   |
| Mn | 0.4  | 0.3   | + 0.1   | 60    | 200 |
| Pb | 0.6  | 0.6   | 0.0     | 19    | 32  |
| Sr | 0.1  | 0.4   | - 0.3   | 100   | 250 |
| Zn | 0.9  | 0.6   | + 0.3   | 141   | 235 |

river water (table 3.18). Output then can be compared with input from rainfall. Maximum system flux occurred for calcium, sodium, and magnesium (table 3.20), and in each of these cases the output was greater than the input. Greater input appears to be associated with greater output from the system.

The difference between input and output is net change. If the quantity is negative, the system is losing material, which is compensated by input from the subsoil through the process of weathering. If the quantity is positive, the system is accumulating the element in the active soil layers or the chemical is being leached into deep soil horizons. Net change was negative for calcium, sodium, magnesium, iron, and strontium and positive for potassium, phosphorus, cobalt, zinc, copper, manganese, and lead (table 3.20). The major conclusion that can be drawn from table 3.20 is that there is a substantial loss of calcium, sodium, and magnesium from the forest and relatively little change in other elements studied. This conclusion is not unexpected since it is well known that calcium, sodium, and magnesium are among the most common exchangeable ions in soils (Lyon and Buckman, 1949).

Aside from the absolute value of the transfers into and out of the forest system, we can also consider these transfers in relation to the total quantity of the element stored in the system. The inventory in the vegetation and active soil layers for each of the 12 elements also is shown in table 3.20. Calcium, potassium, magnesium, phosphorus, and zinc are most abundant in the Tropical Moist forest ecosystem. The standing crop divided by the output through stream discharge is a measure of the rate of turnover of the mineral pool. Turnover time was over 100 years for potassium, phosphorus, strontium, zinc, manganese, and calcium (table 3.20). The turnover times for sodium and iron seem to be unrealistically short. A similar conclusion was made for throughfall and litter fall of these two elements compared to the vegetation inventory. In each case, the result is due to high amounts of these elements in the water samples. Apparently the water samples were contaminated with sodium and iron and for this reason we will not further discuss the cycling of these two elements although data on their abundance will continue to be presented in the tables.

# CHAPTER 8

# A Model of Mineral Cycling in the Tropical Moist Forest Ecosystem

In the preceding sections data have been presented which describe the structural and chemical states of the Tropical Moist forest in eastern Panama. These structural data, together with information on litter fall, movement of water, and animal feeding rates, provide the basis for a description of the cycling of chemicals through the forest. The descriptive technique used here is a deterministic linear model, with the assumption that chemical input to the biotic part of the system equals chemical output from the biota.

For purposes of modelling, the forest was organized into the following components: leaves $(C_1)$, stems $(C_2)$, litter $(C_3)$, soil $(C_4)$, roots $(C_5)$, fruits and flowers $(C_6)$, detritivores $(C_7)$, herbivores $(C_8)$, and carnivores $(C_9)$. Each of these compartments are represented on a block diagram (fig. 1.3). Compartments outside the system of definition were atmosphere $(C_0)$, subsoil $(C_{10})$, and adjacent systems $(C_{11})$ and also are shown on figure 1.3. These compartments do not coincide exactly with those used in the structural analysis of the ecosystem because in the preliminary steps to modeling, certain compartments were discovered to be inconsequential. For example, the distinction between the overstory and understory was dropped because the understory contained a relatively insignificant amount of the chemical inventory and its contribution to the litter could not be separately distinguished. Thus, the appropriate stratal compartments were summed.

The pathways between compartments were designated as $\lambda_{ij}$, where the subscripts denote the donor $(i)$ and receptor compartments $(j)$ respectively. It was assumed that mineral flux in the Tropical Moist forest was in steady state, so that element flux could be described by:

$$\frac{dC_i}{dt} = \sum_{j=1}^{10} (\lambda_{ij} - \lambda_{ji}) = 0$$

where $C_i$ is the concentration in a compartment and $t$ is time. The assumption of steady state is probably justified since the Tropical Moist forest appears to be a mature system without obvious accretion or degradation.

The notation and description of transfer coefficients (kg/ha/yr) for the Tropical Moist forest ecosystems are listed below. The details of the calculation of rates for leaf and twig fall, rainfall, leaf interception, stream runoff, animal feeding, and standing crops of elements within compartments have been described. The seasonal change in the forest is ignored in this preliminary model because of the difficulty of separating the plot differences from season differences.

$\lambda_{0,1}$ Total rainfall striking leaf surfaces. The annual rainfall (1933 millimeters) was converted to kg/ha/yr by dividing 1933 by 10,000 to convert to $kg/cm^2$, then multiplied by $10^6$ to convert to kg/ha. Rain is in area units so annual rain input is in units of 19.3 x $10^6$ kg/ha/yr.

$$\lambda_{0,1} = 19.3 \times 10^6 \text{ kg/ha/yr} \times \text{ppm in rain (table 3.16).}$$

$\lambda_{0,1,3}$ Throughfall to the litter. It was determined that 81 percent of the annual rainfall would strike the leaf surfaces and fall to the forest floor as throughfall.

$$\lambda_{0,1,3} = 19.3 \times 10^6 - 3.67 \times 10^6 \text{ kg/ha/yr} \times \text{ppm in throughfall (table 3.10).}$$

$\lambda_{1,3}$ Leaf contribution to litter. Annual leaf litter fall rate (from table 3.1) so that

$$\lambda_{1,3} = 9,826 \text{ kg/ha/yr} \times \text{ppm overstory leaves (table 2.10).}$$

$\lambda_{1,8}$ Leaf consumption by herbivores. The estimated herbivore consumption rate (0.257 $g/m^2$ day (table 3.13) or 938 kg/ha/yr) × ppm in overstory leaves so that,

$$\lambda_{1,8} = 938 \text{ kg/ha/yr} \times \text{ppm in leaves (overstory leaves in table 2.10).}$$

$\lambda_{2,1}$ Movement from stems to leaves. Equals output from leaves minus input from the atmosphere or

$$\lambda_{2,1} = (\lambda_{1,3} + \lambda_{1,8} + \lambda_{0,1,3}) - \lambda_{0,1}$$

$\lambda_{2,3}$ Branch contribution to litter. Annual branch fall (table 3.1) times the concentration in branches (table 3.4), or

$$\lambda_{2,3} = 1540 \text{ kg/ha/yr} \times \text{ppm in branches.}$$

$\lambda_{2,6}$ Movement from stems to fruits and flowers. Assuming two fruit crops per year, double the standing crop (table 2.3) or

$$\lambda_{2,6} = 278 \text{ kg/ha/yr} \times \text{ppm (average values in table 2.10).}$$

$\lambda_{3,4}$ Infiltration rate of water through the litter to the mineral soil.

Assuming no surface runoff, 100 percent of throughfall $(\lambda_{0,1,3})$ will enter the soil, or

$$\lambda_{3,4} = \lambda_{0,1,3}$$

$\lambda_{3,7}$ Litter consumption by detritivores. Assuming that all contributions to the litter other than throughfall would be consumed by detritivores, or

$$\lambda_{3,7} = \lambda_{1,3} + \lambda_{2,3} + \lambda_{6,3} + \lambda_{8,3} + \lambda_{9,3}$$

$\lambda_{4,5}$ Movement soil to roots. Set equal to $\lambda_{5,2}$.

$\lambda_{4,11}$ Contribution to other systems from the soil was assumed to equal discharge from the surface soil. Discharge (IOCS-FD-59, 1968) for the 259 km² was 203,820 × 10⁶ liters for May to December, or 221,543 × 10⁶ liters for 12 months using the same conversion factors on table 3.17, or 8.59 × 10⁶ kg/ha/yr.

$$\lambda_{4,11} = 8.59 \times 10^6 \text{ ppm in river water (table 3.18)}.$$

$\lambda_{5,2}$ Movement from roots to stems. Equals the output from stems, or

$$\lambda_{5,2} = \lambda_{2,1} + \lambda_{2,3} + \lambda_{2,6}.$$

$\lambda_{6,3}$ Contribution of fruits and flowers to the litter. The difference between frugivore consumption and movement from stems to fruits and flowers, or

$$\lambda_{6,3} = \lambda_{2,6} - \lambda_{6,8}.$$

Table 3.21.   Transfer rates (kg/ha/yr) of

| Element | 0,1 | 0,1,3 | 1,3 | 1,8 | 2,1 | 2,3 | 2,6 | 3,4 | 3,7 | 4,5 |
|---|---|---|---|---|---|---|---|---|---|---|
| P | 0.96 | 0.61 | 7.86 | 2.06 | 9.57 | 0.77 | 0.61 | 0.61 | 11.30 | 10.95 |
| K | 9.49 | 50.02 | 110.05 | 13.41 | 163.99 | 18.63 | 4.84 | 50.02 | 146.93 | 187.46 |
| Ca | 29.25 | 37.51 | 223.05 | 19.89 | 251.20 | 16.63 | 1.75 | 37.51 | 260.32 | 269.58 |
| Mg | 4.86 | 9.75 | 20.63 | 2.53 | 28.05 | 1.54 | 0.61 | 9.75 | 25.31 | 30.20 |
| Na | 30.71 | 24.01 | 1.97 | 0.19 | -4.54 | 0.31 | 0.03 | 24.01 | 2.50 | -4.20 |
| Co | 1.99 | 0.98 | 0.20 | 0.05 | -0.76 | 0.05 | 0.01 | 0.98 | 0.31 | -0.70 |
| Cu | 0.47 | 0.26 | 0.07 | 0.01 | -0.13 | 0.03 | 0.003 | 0.26 | 0.11 | -0.10 |
| Fe | 3.01 | 4.28 | 0.45 | 0.06 | 1.78 | 0.05 | 0.02 | 4.28 | 0.58 | 1.85 |
| Mn | 0.42 | 0.34 | 0.23 | 0.07 | 0.22 | 0.17 | 0.02 | 0.34 | 0.49 | 0.41 |
| Pb | 0.60 | 0.30 | 0.35 | 0.04 | 0.09 | 0.03 | 0.01 | 0.30 | 0.43 | 0.13 |
| Sr | 0.08 | 0.07 | 0.86 | 0.09 | 0.94 | 0.08 | 0.01 | 0.07 | 1.04 | 1.03 |
| Zn | 0.87 | 0.63 | 0.21 | 0.02 | -0.01 | 0.04 | 0.01 | 0.63 | 0.28 | 0.04 |
| Combined | 82.71 | 128.76 | 365.93 | 38.42 | 450.40 | 38.33 | 7.92 | 128.76 | 449.60 | 496.65 |

$\lambda_{6,8}$ Consumption of fruits and flowers by herbivores. The estimated frugivore consumption of 0.028 g/m²/day (table 3.13) or 102 kg/ha/yr × ppm in fruits and flowers in table 2.10:

$$\lambda_{6,8} = 102 \text{ kg/ha/yr} \times \text{ppm in fruits and flowers.}$$

$\lambda_{7,4}$ Contribution to soil by detritivores. Assuming the contribution to detritus is entirely transferred to the soil, then

$$\lambda_{7,4} = \lambda_{3,7}$$

$\lambda_{8,3}$ Contribution of herbivores to litter. Equal to the herbivore intake minus the output of carnivores, or

$$\lambda_{8,3} = (\lambda_{1,8} + \lambda_{6,8}) - \lambda_{8,9}$$

$\lambda_{8,9}$ Consumption of herbivores by carnivores. Estimated carnivore consumption rate was 0.03 g/m² day (table 3.13) or 113 kg/ha/yr, so that

$$\lambda_{8,9} = 113 \text{ kg/ha/yr} \times \text{ppm in herbivores (table 2.15).}$$

$\lambda_{9,3}$ Contribution of carnivores to litter. Assumed to be equal to $\lambda_{8,9}$

$\lambda_{10,4}$ Input to surface from subsoil. Set as the difference between $\lambda_{0,1}$ and $\lambda_{4,11}$.

### RESULTS

There are two main sequences of pathways in the Tropical Moist forest model, $C_1 - C_3 - C_7 - C_4 - C_5 - C_2 - C_1$ and $C_1 - C_8 - C_9$.

12 elements in a Tropical Moist forest.

| Element | 4,11 | 5,2 | 6,3 | 6,8 | 7,4 | 8,3 | 8,9 | 8,3 | 10,4 |
|---|---|---|---|---|---|---|---|---|---|
| P | 0.73 | 10.95 | 0.39 | 0.22 | 11.30 | 2.22 | 0.06 | 0.06 | - 0.23 |
| K | 9.26 | 187.46 | 3.07 | 1.77 | 146.93 | 14.75 | 0.43 | 0.43 | - 0.23 |
| Ca | 163.20 | 269.58 | 1.11 | 0.64 | 260.32 | 18.53 | 1.00 | 1.00 | +133.95 |
| Mg | 43.64 | 30.20 | 0.39 | 0.22 | 25.31 | 2.49 | 0.26 | 0.26 | + 38.78 |
| Na | 92.50 | -4.20 | 0.02 | 0.01 | 2.50 | -0 36 | 0.56 | 0.56 | + 61.79 |
| Co | 0.67 | -0.70 | 0.01 | 0.004 | 0.31 | 0.01 | 0.04 | 0.04 | - 1.32 |
| Cu | 0.36 | -0.10 | 0.002 | 0.001 | 0.11 | -0.01 | 0.02 | 0.02 | - 0.11 |
| Fe | 10.08 | 1.85 | 0.01 | 0.01 | 0.58 | -0.23 | 0.30 | 0.30 | + 7.07 |
| Mn | 0.34 | 0.41 | 0.01 | 0.01 | 0.49 | 0.07 | 0.01 | 0.01 | - 0.08 |
| Pb | 0.58 | 0.13 | 0.01 | 0.003 | 0.43 | 0.02 | 0.02 | 0.02 | - 0.02 |
| Sr | 0.42 | 1.03 | 0.01 | 0.003 | 1.04 | 0.09 | 0.003 | 0.003 | + 0.34 |
| Zn | 0.56 | 0.04 | 0.01 | 0.003 | 0.28 | -0.02 | 0.04 | 0.04 | - 0.31 |
| Combined | 322.34 | 496.65 | 5.04 | 2.894 | 449.60 | 37.56 | 2.743 | 2.743 | +239.63 |

The first sequence represents the vegetation cycle and the second the flow through animal populations. Approximately 90 percent of the flow is through the vegetation and 10 percent through the animal chain (table 3.21). In a comparison of transfer coefficients, the model rules indicate the most significant comparisons are between the outputs from litter $(C_3)$ to detritivores and soil, input to leaves $(C_1)$ from stems $(C_2)$, and inputs to herbivores $(C_8)$ from leaves to fruit. Output from the litter should be numerically large because litter accumulates many different inputs and, therefore, has an integrating function in the system. If output from litter is large, $\lambda_{7.4}$, $\lambda_{4.5}$, $\lambda_{5.2}$ also must be large since there is a direct coupling of litter to stems through soil and roots. The pathway from stems to leaves represents the major output from stems and closes the vegetation sequence $C_1$ - $C_3$ - $C_7$ - $C_4$ - $C_5$ - $C_2$ - $C_1$. The pathway from leaves to herbivores is the main input to the animal sequence of pathways. Transfer coefficients $\lambda_{3.7}$ and $\lambda_{2.1}$ are of special interest since they are at the locus of numerous inputs and/or reflect important biological transfers.

The combined output of 570 kg/ha/yr from $C_3$ (the sum of $\lambda_{3.4}$ + $\lambda_{3.7}$) exceeded the output from all other components (table 3.21). For most elements, except cobalt, copper, iron, zinc, and sodium, $\lambda_{3.7}$ was numerically greater than $\lambda_{3.4}$. That is, more cobalt, copper and the other elements above moved in throughfall water to and through the litter than moved in litter fall of leaves.

Transfer from stems to leaves $(\lambda_{2.1})$ was also numerically large (table 3.21) in comparison with the other transfers in the system. Output to leaves from stems was greater by a factor of 10 than the other outputs from stems to fruits and flowers and to litter through branch fall.

The flow to the herbivores $(\lambda_{1.8}$ and $\lambda_{6.8})$ averaged a little more than 10 percent of the input to the leaves $(\lambda_{2.6})$. Variation between elements depends upon variability in element content of the leaves. As presented in chapter 6, the animal intake was estimated from information of density, biomass, and weight-specific food consumption. Other investigators have measured tree foliage utilization directly by determining the area of leaves consumed by insects. While this technique ignores consumption of wood, fruit, and seeds, and the effect of leaf growth on the area eaten, it provides comparative data to determine if our estimates of flow to herbivores are realistic. Estimates of leaf consumption as percent standing crop of leaves are 1.4 to 11.7 percent for north temperate forests (Bray, 1961 and Lomnicki et al., 1965), and 5.6 percent for Tropical Montane forest (de la Cruz, 1964). Leaf consumption as percent of above-ground production is estimated to range from 0.5 to 8.5 percent (Bray, 1961; Lomnicki et al., 1965; Misra, 1968; and Reichle and Crossley, 1967). Clearly our independent computation of herbivore intake resulted in an estimate that was reasonable compared to other forest environments.

Input to the forest system is equal to $\lambda_{0.1}$ and $\lambda_{10.4}$. Input from rain was very much less than the transfer through the vegetation sequence of pathways, but was greater than transfer to animals. Input from the subsoil was positive for calcium, magnesium, sodium, iron, and strontium and negative for the other elements, indicating large inputs of calcium, magnesium, sodium, and iron to balance the losses to the system in river discharge. Daily discharge from the Sabana watershed of 25,911 hectares ranged from 24 to 750 cubic feet per second (IOCS-FC-59, 1968) and carried a large load of calcium, sodium, and magnesium (table 3.18).

The input from rainfall and weathering and the output from stream discharge can be compared with the biological cycle of elements by comparing $\lambda_{4.11}$, the system output, with $\lambda_{5.2}$ the uptake of stems from roots or with the litter fall $\lambda_{1.3}$ (table 3.21). The percentage transfer through the biological cycle compared to the geological cycle is in the following order $P > K > Mn = Sr > Ca$. For the other elements the percentage transfer in geological cycle is greater than that in the biological cycle.

One of the most interesting comparisons of transfer rates for individual elements is the movement of minerals in water through rainwash and in litter. Both processes operate because of gravity and together they represent the major downward movement of minerals in terrestrial systems. Rainwash as a percentage of litter fall (table 3.22) was less than 100 percent for all elements except copper, cobalt, zinc, and manganese. Relatively little phosphorus, strontium, or calcium, or magnesium moved through the rainwash. Another important comparison concerns the movement by gravity in litter fall and the opposite movement against gravity via biological uptake. Uptake is greater for all elements except copper, cobalt, zinc, and lead (table 3.22). For these four elements the calculations resulted in negative transfer functions from roots to stems.

Another way to examine the dynamics of these tropical forest ecosystems is through comparison of compartment turnover, which is the ratio of the standing crop of an element in a compartment to the sum of outputs from the compartment. The turnover is expressed in years and can be termed turnover time. Turnover time varies widely for a given element across compartments (table 3.23), and among elements in a compartment. In general, turnover is most rapid for compartments $C_7$ (detritivores), $C_8$ (herbivores) and $C_9$ (carnivores). It is longest for $C_2$ (stems) and $C_4$(soils). Turnover time for the leaf compartment is consistently about one year or less. Litter and fruits and flowers also turnover in less than one year. Turnover time of animal compartments is almost always less than a few days.

Table 3.22.   Comparison of the rate of movement of minerals in rainwash ($\lambda_{0,1,3}$) vs. litter fall ($\lambda_{1,3}$) and in biological uptake ($\lambda_{5,2}$) vs. litter fall ( $\lambda_{1,3}$) in Tropical Moist forest ( × 100).

| Element | Rainwash/litter fall | Uptake/litter fall |
|---|---|---|
| Phosphorus | 8 | 139 |
| Potassium | 46 | 170 |
| Calcium | 17 | 121 |
| Magnesium | 47 | 146 |
| Cobalt | 490 | - 35 |
| Copper | 371 | -143 |
| Manganese | 148 | 178 |
| Lead | 86 | - 37 |
| Strontium | 8 | 120 |
| Zinc | 300 | - 2 |

This chapter has focused on the biotic portion of the Tropical Moist forest ecosystems. We have shown that there are two major sequences of pathways, that of the vegetation and active soil layers (to 30 cm) and that of the herbivores and carnivores. The assumptions of the model required input from the atmosphere and weathering to equal output from the active soil layer, so that the vegetation and animal cycles were in equilibrium. It seems reasonable that the system would be in equilibrium with its environment since the forest probably has been present in Darien Province at least for several hundred years. Nevertheless, there may be leaching of minerals downward through the active soil layer at rates greater than input from the atmosphere, as has been reported for Douglas fir forests by Cole et al. (1968). These losses may be countered by inputs from the subsoil. If

Table 3.23.  Turnover in years for 12 elements in compartments of the Tropical Moist forest.

| Element | Atmosphere $C_0$ | Leaves $C_1$ | Stems $C_2$ | Litter $C_3$ | Soil $C_4$ | Roots $C_5$ | Fruits $C_6$ | Detritivores $C_7$ | Herbivores $C_8$ | Carnivores $C_9$ |
|---|---|---|---|---|---|---|---|---|---|---|
| P  | 1 | 1.6 | 11.7  | 1.2 | 1.9   | 0.5  | 0.3 | 0.003 | 0.004 | 0.02 |
| K  | 1 | 0.8 | 15.2  | 0.2 | 1.8   | 0.4  | 0.3 | 0.001 | 0.003 | 0.02 |
| Ca | 1 | 0.9 | 12.4  | 1.1 | 51.2  | 0.8  | 0.3 | 0.002 | 0.007 | 0.10 |
| Mg | 1 | 0.9 | 11.8  | 0.6 | 30.6  | 0.9  | 0.3 | 0.006 | 0.001 | 0.5 |
| Na | 1 | 0.9 | 11.4  | 0.1 | 12.7  | 0.7  | * | 0.10 | 0.35 | 0.04 |
| Co | 1 | 2.0 | 15.4  | 0.4 | 233.7 | 0.3  | * | * | 0.20 | 0.3 |
| Cu | 1 | 1.3 | 14.0  | 0.3 | 25.0  | 1.0  | * | * | * | * |
| Fe | 1 | 0.4 | 4.6   | 1.0 | 2.2   | 0.3  | * | 0.14 | 0.57 | 0.1 |
| Mn | 1 | 1.7 | 66.3  | 1.4 | 40.7  | 1.2  | * | 0.02 | * | * |
| Pb | 1 | 1.3 | 56.9  | 1.1 | 14.1  | 3.1  | * | 0.02 | * | * |
| Sr | 1 | 1.1 | 16.9  | 1.4 | 54.1  | 1.1  | * | * | * | * |
| Zn | 1 | 0.9 | 167.5 | 0.4 | 222.5 | 10.0 | * | 0.04 | 0.50 | * |

* Only trace quantities in the standing crop.

our assumption of equilibrium is incorrect then the transfer rates will be greater than those shown in table 3.21 especially those connecting the animals and vegetation. At this point we feel that it is reasonable to assume that the biotic part of the system is in equilibrium.

### SEASONAL DIFFERENCES IN TRANSFER RATES

The annual model does not take into account the single most important event in the life of the Tropical Moist forest—the alternation of wet and dry seasons. How does the variation in rainfall, with its effect on throughfall, litter fall, and stream discharge, affect the movement of chemicals in the forest? What is the consequence of a lengthening or shortening of the dry season?

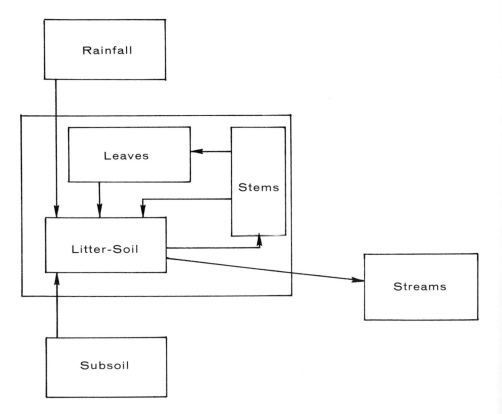

Figure 3.6.   Seasonal model of mineral cycling in the Tropical Moist forest.

Table 3.24.   Comparison of dry and wet season system inputs and
outputs (kg/ha) of elements in the Tropical Moist forest.
The dry season is from January 1 to April 30.

| Elements | Input in rain | | Output in stream discharge | | Output in subsoil* | | | |
|---|---|---|---|---|---|---|---|---|
| | Dry | Wet | Dry | Wet | Dry | | Wet | |
| P | .05 | .91 | .06 | .67 | - | .01 | + | .24 |
| K | .51 | 9.00 | .73 | 8.51 | - | .22 | + | .49 |
| Ca | 1.56 | 27.73 | 12.95 | 150.25 | -11.39 | | -122.52 | |
| Mg | .26 | 4.61 | 3.45 | 40.08 | - 3.19 | | - 35.47 | |
| Na | 2.08 | 28.63 | 7.34 | 85.14 | - 5.26 | | - 56.51 | |
| Co | .11 | 1.89 | .05 | .62 | + 0.06 | | + 1.27 | |
| Cu | .03 | .45 | .03 | .33 | 0.00 | | + 12 | |
| Fe | .16 | 2.85 | .80 | 9.26 | - 0.64 | | - 6.41 | |
| Mn | .02 | .40 | .02 | .32 | 0.00 | | + .08 | |
| Sr | .01 | .08 | .03 | .39 | - .02 | | - .31 | |
| Zn | .05 | .82 | .04 | .51 | + .01 | | + .31 | |
| Total | 4.84 | 77.37 | 25.50 | 296.08 | ——— | | ——— | |

* A negative sign indicates a contribution from the subsoil to the top soil. A positive
sign indicates leaching from the top soil to the subsoil.

These questions can be partly answered with a simplified seasonal model (fig. 3.6) which focuses on the three inputs to litter, the
leaf fall, branch fall and throughfall, and uptake by stems, as well as
the system inputs and outputs. The average dry season has a length
of 120 days, and during the study 103 mm of rain fell within this
period. Since 1,830 mm fell during the 245 day wet season, the input
of chemicals by rain differed greatly between seasons (table 3.24).
Stream discharge is also related to rainfall but the difference between
seasons is less extreme because of the effect of water storage in the soil.
During the dry season there was a general net flow from the subsoil to
the top soil since discharge was greater than rain input. In contrast,
during the wet season there was a strong leaching effect, with phosphorus, potassium, cobalt, copper, manganese, and zinc moving from
the topsoil to the subsoil (table 3.24). Calcium, magnesium, sodium,
and strontium, however, continued to be exported from the forest
system.

Within the forest, litter fall increases during the dry season yet the difference in season length results in an absolute larger quantity of leaves falling during the wet season. Wet season input is 5,709 kg/ha as contrasted to 3,984 kg/ha during the dry season. Branch fall during the dry season is greater (864 kg/ha) than during the wet season (564 kg/ha). Throughfall is greater during the wet (1,482 mm) compared to the dry (83 mm) season. As a consequence of the differences in inputs and the change in concentration of leaves and stems seasonally, there is considerable difference in season length results in an absolute larger quantity of leaves falling during the wet season. Wet season input is 5,790 kg/ha as contrasted to 3,984 kg/ha during the dry season. Branch fall during the dry season is greater (864 kg/ha) than during the wet season (564 kg/ha). Throughfall is greater during the wet (1,482 mm) compared to the dry (83 mm) season. As a consequence of the differences in inputs and the change in concentration of leaves and stems seasonally, there is considerable difference in the contribution of the three inputs to the litter (table 3.25). For example, leaf fall total input during the dry season is less than one-half that during the wet season,

Table 3.25.   Comparison of seasonal differences in cycling rates (kg/ha) in Tropical Moist forest. The dry season is from January 1 to April 30.

| Element | Seasonal leaf fall | | Seasonal branch fall | | Seasonal throughfall | | Daily stem uptake $\times 10^2$ | |
|---|---|---|---|---|---|---|---|---|
| | Dry | Wet | Dry | Wet | Dry | Wet | Dry | Wet |
| P | 3.19 | 12.56 | 0.25 | 0.28 | 0.03 | 0.58 | 33 | 55 |
| K | 44.62 | 81.64 | 4.67 | 6.82 | 2.66 | 47.42 | 43 | 52 |
| Ca | 90.44 | 121.03 | 9.68 | 6.09 | 1.99 | 35.57 | 88 | 55 |
| Mg | 8.37 | 15.41 | 1.21 | 0.56 | 0.52 | 9.25 | 8 | 8 |
| Na | 0.80 | 1.14 | 0.09 | 0.11 | 1.27 | 22.76 | 0.07 | - 1.9 |
| Co | 0.08 | 0.32 | 0.03 | 0.02 | 0.05 | 0.92 | 0.1 | - .3 |
| Cu | 0.03 | 0.03 | 0.01 | 0.01 | 0.02 | 0.25 | 0.02 | - .07 |
| Fe | 0.18 | 0.36 | 0.02 | 0.02 | 0.23 | 4.05 | 0.2 | 0.6 |
| Mn | 0.09 | 0.43 | 0.05 | 0.06 | 0.02 | 0.33 | 0.1 | 0.2 |
| Sr | 0.35 | 0.54 | 0.06 | 0.03 | 0.01 | 0.07 | 0.3 | 0.2 |
| Zn | 0.08 | 0.11 | 0.02 | 0.01 | 0.03 | 0.59 | 0.07 | - .04 |
| Total | 148.23 | 233.57 | 16.09 | 14.01 | 6.83 | 121.79 | 142.86 | 118.69 |

while branch input is slightly greater during the dry season. In contrast, throughfall is about 20 times greater during the wet season. The daily rate of uptake by the stems is greater during the wet season for phos-phosphorus  and potassium but not for calcium (table 3.25). These differences between elements are due to the relative concentrations and solubilities in vegetative material. Calcium which is a component of the structural material of the trees is associated mainly with leaves and stems, while potassium is highly water mobile and relative to its inventory appears in large quantities in the throughfall.

From these data, we can speculate that extension of the dry season for one or more months will decrease the rain input, stream discharge, and throughfall. Assuming that litter and branch fall continue at dry season rates, then the mineral cycle within the forest will decline at approximately the rate in table 3.25. The effect will be especially strong for elements such as potassium.

# CHAPTER 9

# Analog Simulation
# of the Potassium Cycle[1]

In order to develop the conceptual model described in the previous chapter it was necessary to make several assumptions concerning accuracy of the data and the relationship between components. It is possible to test the sensitivity of the model by altering the values of the transfers in a regular way and thereby determine the significance of certain types of assumptions on the overall behavior of the system. Two ways to accomplish this test are: to remove the forcing from rain input and allow the model to decay over time (free behavior), and to force the system and examine the behavior of the compartments over time.

Potassium was chosen for the analog simulation study since the potassium data were substantiated by other studies; potassium could be consistently determined in all parts of the system, and the potassium cycle seemed straight-forward, following the assumptions set up for the model. The analog computer diagram (fig. 3.7) was derived directly from the model shown in fig. 1.3. The scaling equations are listed in table 3.26. The model was a linear, constant coefficient, first order differential equation model.

## FREE BEHAVIOR OF THE MODEL

When the forcing of rainfall is removed, key system components behave as shown in fig. 3.8. Leaves receive the input; when the input terminates, the standing crop of potassium in leaves initially increases slightly and then declines at a constant rate. The amount of potassium in litter also increases initially, but the increase is much larger and is prolonged. On the time scale of the model, litter potassium increases progressively for about 10 years before a decline, at a constant rate.

[1]The analog simulation was performed by Dr. Clayton Gist, Institute of Ecology, University of Georgia.

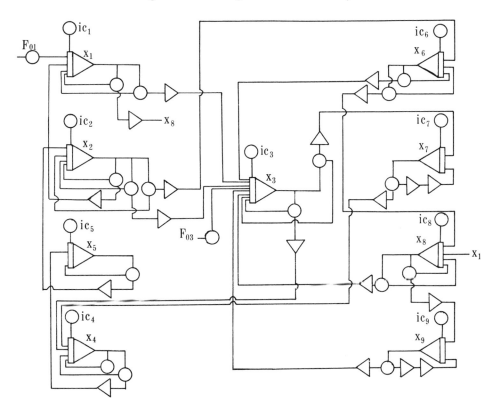

Figure 3.7. Computer diagram for the potassium simulation experiment. Initial conditions are indicated as $ic_n$, compartments by $X_n$, and forcings as $F_n$. The symbols indicate integrators (triangle and rectangle combined), pots (circles) and inverters (triangles).

The behavior of the litter compartment is explained in part by the fact that it accumulates inputs from all other above-ground compartments (litter has five inputs and two outputs) and, therefore, decline in these compartments results in an increase in litter before loss from the system.

In contrast to litter, the soils and roots exhibit a decline in initial level of potassium, followed by an increase and then a constant decrease. The initial decrease in potassium is due to the loss of the element from the soil in stream discharge. This loss continues unabated

Table 3.26.    Scaling equations used in analog modelling.

$X_1$    $.5\dot{X}_1 = [.5/3.3][3.3F_{01}] + [.5/.02](6.2 \times 10^{-2})[.02X_2]$
$- \{[1](1.21 \times 10^{-1})|(10^3X_8) + [1](8.8 \times 10^{-1})(2.0X_3)\}$

$X_2$    $.02\dot{X}_2 = [.02/1](2.46)[X_5]$
$- \{[1](6.5 \times 10^{-3})[2.0X_3] + [1](1.7 \times 10^{-3})(2X_6) + [1](6.2 \times 10^{-2})[.5X_1]\}$

$X_3$    $2.0\dot{X}_3 = [2.0/.1][.1F_{03}] + [2./.5](8.8 \times 10^{-1}/)[.5X_1] + [2.0/10^3](3.5 \times 10^1)[10^3X_9]$
$+ [2.0/10^3](3.7 \times 10^2)[10^3X_8] + [2.0/2](1.85 \times 10^{-1})[2X_6]$
$+ [2.0/.02](6.5 \times 10^{-3})[.02X_3] - \{[1](4.08)[70X_7] + [1](1.28)[.2X_4]\}$

$X_4$    $.2\dot{X}_4 = [2./2.0](1.28)[2.0X_3] + [.2/70](1.27 \times 10^3)[70X_7]$
$- \{[1](2.69 \times 10^{-2})[X.] + [1](5.67 \times 10^{-1})[1X_5]\}$

$X_5$    $.1\dot{X}_5 = [1/.2](5.67 \times 10^{-1})[.2X_4] - \{[1](2.46)[.02X_3]\}$

$X_6$    $2\dot{X}_6 = [2/.02](1.7 \times 10^{-3})[.02X_2]$
$- \{[1](1.85 \times 10^{-1})[2.X_3] + [1](1.85 \times 10^{-1})[10^3X_8]\}$

$X_7$    $70\dot{X}_7 = [70/2.0](4.08)[2.0X_3] - \{[1](1.27 \times 10^3)[.2X_4]\}$

$X_8$    $10^3\dot{X}_8 = [10^3/2](1.85 \times 10^{-1})[2X_6] + [10^3/.5](1.21 \times 10^{-1})[.5X_1]$
$- \{[1](3.37 \times 10^2)[2.0X_3] + [1](9.26)[10^3X_9]\}$

$X_9$    $10^3\dot{X}_9 = [10^3/10^3](9.26)[10^3X_8] - [1](3.5 \times 10^1)[2.0X_3]$

and drains the compartment, then as litter accumulates potassium, the soil potassium content also increases. Roots follow this same pattern since they derive their potassium solely from the soil compartment.

The potassium content of stems does not exhibit fluctuation in content. Rather stem potassium declines progressively. It requires more than 250 years for one-half of the original pool of potassium to be lost. This rate of decrease may be unrealistic biologically since decomposition would be expected to increase under tropical conditions. Rather it reflects the model assumption and design, in which the change in the potassium stem pool is governed by the transfer rate from stems to leaves, litter and other compartments. The data clearly show how the other components are dominated by the rate of decline in the stems; for example, all decrease curves after the initial fluctuation have the same slope as the curve for stems. We conclude that stems have a strong damping effect on the system through the large size of the mineral pool and the relatively low rate of transfer into and out of the pool.

The potassium content in herbivores shows a rapid decline initially before settling into the progressive decline dominated by the stem

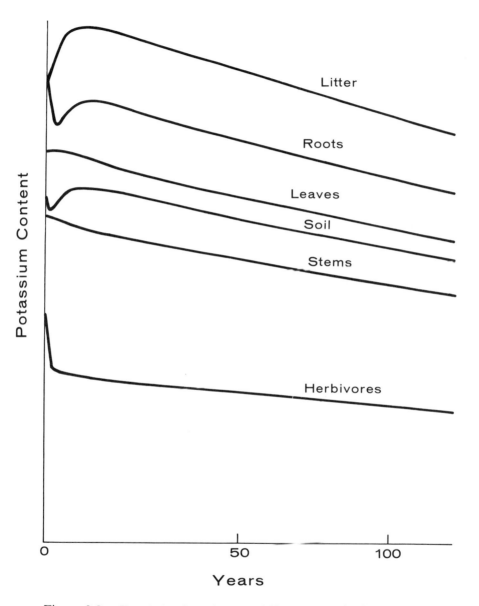

Figure 3.8.   Free behavior of selected Tropical Moist forest components when forcing of potassium is removed.

dynamics. The herbivores do not reflect the slight increase in potassium content of leaves in the first period after the forcing is stopped. The herbivores have a relatively small transfer from the leaves and a small compartment size (thus little storage capacity) and these characteristics result in a high degree of sensitivity to changes in other compartments. This type of linear model reveals several important characteristics of animal groups. If, as in the case of the herbivores, the amount of transfer between food and animal compartments is small and the standing crop of animals is low, the animals are highly sensitive to changes in the food compartments. If, similar to the detritivores, the transfer between the animal systems is large and the mass of animals is small, the animals are effective as a control and the system is sensitive to the animals. In another test of this model on the digital computer using CSMP, it was found that the detritivore compartment had to be increased up to 12 times its size for potassium before the model stabilized. In other words, the detritivores were underestimated by about 12 times and the sensitivity of the system to this compartment was such that very small differences in the mass of decomposers caused an imbalance in the transfers.

### FORCED BEHAVIOR OF THE MODEL

An alternate test of the system is to increase the forcing on the leaf compartment. In this test the leaf compartment was forced with a square wave generator to about six units above its equilibrium level (fig. 3.9). The object was to observe the change in height and time lag of the other system compartments.

Litter showed the greatest effect of forcing the leaf compartment because the major loss from the leaf compartment was to the litter and the major input to the litter was from the leaf compartment. The litter equilibrium level was increased about 3½ units and lagged one time unit after the peak of the leaf increase. Soil and roots increased about two units and lagged two time units. Herbivores increased one and one-half units and showed no time lag in response verifying the previous observation on the small storage capacity of herbivores. Stems showed no response at all to a change in leaf potassium level.

These observations support those described under the free behavior of the model. Herbivores are very responsive to changes in leaves but the capacity for change in size of the mineral pool is limited. Stems are unresponsive and act as a damping factor in the system. Litter, soil, and roots are responsive to changes in forcing. Increase in the

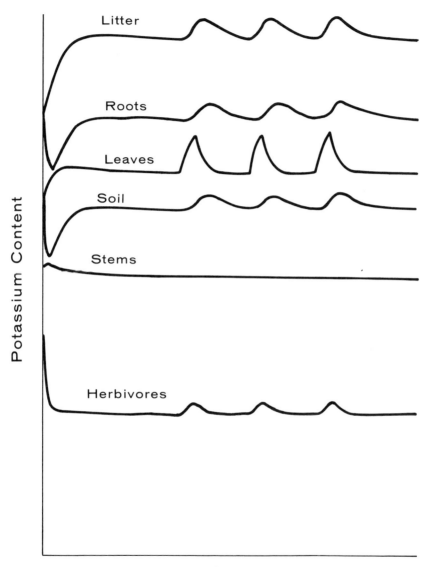

Figure 3.9.   Forced behavior of potassium in Tropical Moist forest
components, after forcing the leaf compartment.

size of the pool in either soil or roots would be expected to decrease the magnitude of their response.

Regarding modeling assumptions, these experiments suggest that errors in estimation of herbivore biomass and inputs may not be as important as errors in detritivore estimates. Further, stems are of major importance in controlling system behavior. Stems can be rather accurately sampled and we are relatively confident of our estimates of stem biomass in this study. On the other hand, several pieces of evidence cited here and earlier suggest that detritivores were under-estimated. In mineral cycling, the detritivore component could function as an important control in the cycling sequence.

# IV
# COMPARISON OF TROPICAL FORESTS AND MINERAL CYCLING

*On the whole it may be said the living organisms are composed of comparatively rare elements. We are, indeed, earth born, but yet not altogether common clay.*

ALFRED J. LOTKA

The Tropical Moist forest is an example of only one of many types of tropical forest. Within the Darien Province there are at least three other forest types, and each of these types is quite different from the Tropical Moist forest. The first set of comparisons will focus on these Darien forests occurring within the same local area. Later comparisons will consider examples of other types within the tropical biome or region. Data on these more varied types of tropical forests will be examined to describe the dimensions of the structure and function of tropical forest generally, as well as to clarify the unique role of Tropical Moist forest.

# CHAPTER 10

# *The Forests of Darien*

In the Darien Province of the Republic of Panama there are a variety of forests. We have described the Tropical Moist forest which occupies about 75 percent of the area. In addition to this semideciduous type, there are Mangrove forests on the Atlantic and Pacific coasts, Riverine forest on the alluvial flats along the major rivers, and Premontane Wet forest on the highest portions of the coastal mountain ranges. Relatively little cultivated land is present. The approximate area of each type of vegetation is shown in table 4.1. In this study each forest type was sampled by one-quarter hectare square plots using methods similar to those described for Tropical Moist forest. Thus, the structural analyses of the Darien forests are comparable with each other.

Table 4.1.   Approximate areal extent of vegetation types in the eastern Panama study area as determined from aerial photos and map tracings. The study area is limited by the dotted lines on figure 2.1.

| Vegetation types | Area (km²) | Percent |
|---|---|---|
| Tropical Moist forest | 3139 | 75.4 |
| Premontane Wet forest | 561 | 13.5 |
| Riverine forest | 107 | 2.6 |
| Mangrove forest | | |
|   Atlantic | 24 | 0.6 |
|   Pacific | 260 | 6.2 |
| Second-growth and cultivated land | 73 | 1.7 |
|     Total | 4164 | 100.0 |

DESCRIPTION OF THE FOREST COMMUNITIES

Pacific coast mangrove forests cover approximately 6 percent of the total land area (table 4.1). This type includes the minor brackish water communities such as black mangrove *(Avicennia germinans),* mora *(Mora oleifera)* and castaño *(Montrichardia arborescens)* which develop along a salinity gradient in the rivers and estuaries in eastern Panama (Mayo, 1965). Mangrove occurs in the most saline condition and castaño in the least saline. The major Pacific brackish water community is the red mangrove *(Rhizophora brevistyla),* which are among the tallest mangroves in the world (30-40 meters, according to West, 1956; Cuatrecasas, 1958; and Mayo, 1965). Pure stands of this evergreen forest (fig. 4.1) form along the open deltaic coast and rivers. Where the tides fluctuate between about two and seven meters twice daily, a dense network of prop roots rises two to three meters or more above the soil surface (fig. 4.2). Allochthonous organic debris is continually trapped in this network of roots and is added to the organically rich soil. The soil is a deep mud of fine-grained alluvium interlaced with many roots and rootlets (Mayo, 1965; West, 1956). The system is virtually monospecific, dominated by tall, slender, widely-spaced trees and is practically devoid of understory vegetation. A list of species of plants observed in this forest by James Duke is in appendix table 2. The up-curving branching pattern of the trees and their wide spacing create a relatively open canopy with only 50 percent of the surface covered by tree leaves and branches. Although vines and epiphyllae are rare, some epiphytes are present in the upper canopy. Tree bark usually supports a thin layer of nonvascular plants, and dense mats of algae cover the prop roots below the high-tide water line. These characteristics are compared with the other Darien forests in table 4.2.

Caribbean coastal islands, inlets and bays support narrow bands of a short-growing form of the red mangrove *(Rhizophora mangle)* and the less abundant but taller white mangrove *(Laguncularia racemosa).* Since the land involved is only about one percent of the total area studied (table 4.1), this minor community was not included in the study. Stands are generally monospecific, less than five meters tall with short prop roots. Inundation rarely exceeds one meter. For detailed information concerning this type, see Golley et al. (1962).

The Riverine forest represents a minor portion of the total land area in eastern Panama, but it is physiognomically and floristically distinct and of major economic importance (Lamb, 1953). This forest is restricted to the low riparian areas and is adapted to soil conditions

Figure 4.1.   Aerial view of the Red Mangrove forest site in eastern Panama, after some of the trees have been harvested. (Photograph by Peter McGrath.)

Figure 4.2. Interior view of the Red Mangrove forest site in eastern Panama. (Photograph by the author.)

Table 4.2.  Characteristic features of four Panamanian tropical forests.

| Feature | Tropical Moist | Mangrove | Riverine | Premontane Wet |
|---|---|---|---|---|
| Palms | Common to Abundant in sub-canopy | Absent | Common in understory | Common in canopy and understory |
| Vines | Common in sub-canopy and canopy | Absent | Common in canopy | Common in canopy |
| Epiphytes | Uncommon | Rare | Common | Abundant |
| Heliconia-like plants | Abundant | Absent | Common | Rare |
| *Cecropia* spp. | Common | Absent | Rare | Rare |
| Deciduous trees | Many in canopy and sub-canopy (semi-deciduous) | Absent (Evergreen) | Absent (Evergreen) | Few in canopy (Virtually Evergreen) |
| Litter fall rate | Seasonal | Constant | Constant | Variable |
| Plank buttressed trees | Rare | Absent | Rare | Common |
| Stilt and prop roots | Rare | Abundant | Absent | Common |
| Relative species diversity | Medium | Low (Virtually monospecific) | Low-medium | High |
| Number of plant species observed (see appendix tables) | 399 | 19 | 218 | 382 |

which result from frequent flooding, (figs. 4.3 and 4.4). Inundation is tied closely to the intensity and duration of rainfall throughout the watershed.

Like the Mangrove forest, the Riverine forest is relatively simple in composition and contains many tall canopy trees (fig. 4.4). See appendix table 1 for the list of species occuring in this forest. The canopy, dominated by cativo *(Prioria copaifera),* has a height of 50-55 meters. Evergreen trees produce a fairly even leaf fall throughout the year. Signs of leaf thinning or synchronous leaf fall are not noticeable, although fruit production of cativo is seasonal, occuring once and possibly twice annually. Litter accumulates during the dry season, but is removed or redistributed by flood waters.

The lower story, separated from the canopy by a vertical area nearly void of vegetation (fig. 4.4), is rich in numbers and species of small palms, but contains few other species. The species composition of the understory is strongly influenced by the topography and is evidently related to the frequency and severity of inundation.

Figure 4.3. Aerial view of Riverine forest on the Chucunaque River floodplain in eastern Panama. Most of the canopy trees are cativo (*Prioria copaifera*). (Photograph by Peter McGrath.)

Figure 4.4.　Inundated Riverine forest in eastern Panama. Tall trees are cativo (*Prioria copaifera*). (Photograph by Peter McGrath.)

Woody and herbaceous vines are encountered regularly and are generally associated with areas of recent disturbance. Fleshy climbing vines commonly grow on the boles of the tall canopy trees (fig. 4.4). Epiphytes, although not abundant, occur in the canopy and sub-canopy. Nonvascular stem epiphytes are poorly developed on the dominant cativo trees and are absent on most palms.

The Premontane Wet forest is limited to the mountainous regions ranging in elevation from 250 to 600 meters. The cordillera forming the continental divide in eastern Panama and western Colombia has maximum elevations occasionally exceeding 600 meters. The main axis of the range extends east-west with many discordant, randomly oriented mountains and valleys. Relief is steep, with slopes frequently exceeding 35 degrees. Geologically, the mountain range is a partially eroded, igneous mass of gabbro, diabase, and other basalts. Bright red latosolic soils, rich in iron, cover the majority of the mountainous regions. The soil is well structured, resulting in good drainage and minimal surface runoff. Located primarily on the San Blas cordillera, land of this vegetation type covers approximately 13.5 percent of the total area studied in Panama (table 4.1).

Climatic conditions in the mountains differ from those in other parts of the Panama isthmus. Rainfall distributed during about nine months exceeds 250 cm annually. Precipitation is primarily orographic origi-nating from moisture-laden Caribbean air masses. Low clouds and cool temperatures maintain a relatively humid condition throughout the year, even in the dry season.

A short, dense forest with a fairly even closed canopy and many sub-canopy palms characterizes the vegetation (fig. 4.5). Canopy trees rarely exceed 30 meters in height and are of deliquescent growth form, but not umbrella-like as at the lower elevations. Canopy and sub-canopy trees are slender (fig. 4.6) and are regularly less than one meter in diameter at breast height. Plant buttresses and stilt roots are common among canopy and subcanopy trees and palms.

The understory contains many species and individuals of palms and numerous other trees and tree ferns, yet is relatively open. The year-round canopy cover and the resultant low levels of incident radiation near the ground probably account for the openness. The species ob-served in this forest are listed in appendix table 3.

Vegetation is evergreen with only a few canopy species being de-ciduous during the dry season. Lianas are common. Vascular and non-vascular epiphytes occur frequently in the canopy and on the boles of trees and palms. Colonies of epiphyllae are well developed on sub-

Figure 4.5.   Aerial view of Premontane Wet forest in eastern Panama. (Photograph by Peter McGrath.)

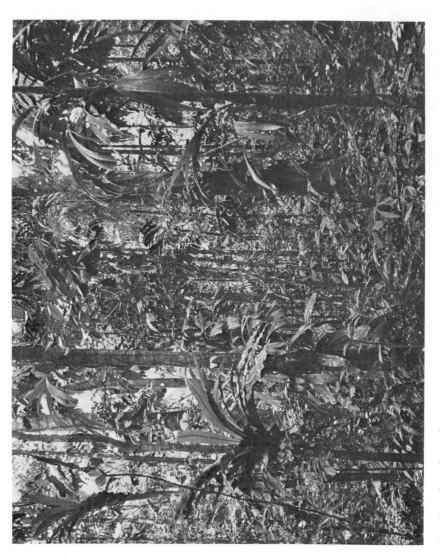

Figure 4.6.  Interior view of Premontane Wet forest in eastern Panama. (Photograph by Peter McGrath.)

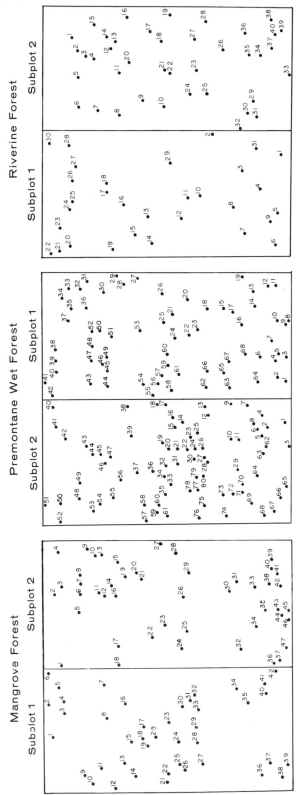

Figure 4.7. Distribution of stems in Mangrove, Premontane Wet, and Riverine forests. The points present trees over 10 cm DBH, on one-fourth hectare. See appendix tables 6–9 for the DBH and names of trees corresponding to the numbers on the map.

canopy and understory leaves. Flowers and fruits are produced throughout the year.

The distribution of stems over 10 cm DBH is shown on maps of the one-quarter hectare plot in each forest (fig. 4.7). The actual diameters and names of the larger trees are listed in appendix tables 6-9. Clearly the Premontane Wet site had the greatest density of stems and the Riverine forest the least. In all three cases a chi-square test of goodness of fit to the Poisson distribution showed that the stems, like those in the Tropical Moist forest, were not distributed over the plot at random. In the Mangrove and Premontane Wet, distinct clumps of stems occur throughout the plot, while the Riverine forest stems appear more evenly distributed.

Table 4.3.    Density and basal area of plants on one-quarter hectare in four Panamanian tropical forests.

| Forest | Number trees other than palms | Number palms | Number vines | Total density | Total m² basal area | Average DBH cm | Maximum tree height m |
|---|---|---|---|---|---|---|---|
| Tropical Moist | 1512 | 142 | 2856 | 4754 | 11.32 | 1.77 | 40 |
| Mangrove | 178 | 0 | 0 | 178 | 3.39 | 12.17 | 41 |
| Riverine | 948 | 381 | 3083 | 4557 | 14.91 | 1.45 | 49 |
| Premontane Wet | 1726 | 898 | 3880 | 6535 | 8.27 | 1.65 | 31 |

The actual number of stems was much greater in the Premontane Wet forest (table 4.3). In this type, trees, palms, and vines were more numerous although the basal area and average diameter were less than Tropical Moist forest. Mangrove forest had the lowest density of stems, and basal area, yet the largest average DBH. In the Mangrove forest there are relatively few large trees. The Premontane Wet forest is of lower maximum height than the three other forests (table 4.3). The maximum height of 49 meters in the Riverine forest is not unusual for tropical forests. Cain and Castro (1959) list maximum heights of tropical forest trees up to 84 meters in Borneo.

Table 4.4.  Representative tree densities of tropical forests.

| Forest types | Location | Number trees/ha ≥10 cm DBH | Number trees/ha ≥20 cm DBH | Source |
|---|---|---|---|---|
| Tropical Moist | Panama | 480 | 168 | Present study |
| | Panama | 424 | 84 | Present study |
| | Panama | 313 | —— | Sexton et al., 1963 |
| Red Mangrove | Panama | 407 | —— | Holdridge, 1964 |
| | Panama | 348 | 112 | Present study |
| | Panama | 313 | —— | Mayo, 1965 |
| Riverine and Swamp | Panama | 590 | —— | Sexton et al., 1963 |
| | Panama | 473 | —— | Mayo, 1965 |
| | Panama | 408 | —— | Holdridge, 1964 |
| | Southern Nigeria | 360 | 241 | Richards, 1939 |
| | Panama | 252 | 144 | Present study |
| | British Guiana | 310 | – – | Richards, 1952 |
| Montane | Ecuador | 495 | 301 | Grubb et al., 1963 |
| | Panama | 592 | 192 | Present study |
| Other types | | | | |
| Morabukea | British Guiana | 309 | —— | Richards, 1952 |
| Mixed | British Guiana | 432 | —— | Richards, 1952 |
| Greenheart | British Guiana | 519 | —— | Richards, 1952 |
| Wallaba | British Guiana | 617 | —— | Richards, 1952 |
| *Cavanillesia*—Mixed | Panama | 237 | —— | Mayo, 1965 |
| Dipterocarp | Thailand | 581 | —— | Ogawa et al., 1965a |
| Monsoon | Thailand | 494 | —— | Ogawa et al, 1965a |
| Lowland | Dominica | 365 | 247 | Grubb et al., 1963 |
| Lowland | Tobago | 398 | 258 | Grubb et al., 1963 |
| Lowland | Nigeria | 420 | 237 | Grubb et al., 1963 |
| Primary Mixed | Southern Nigeria | 441 | 209 | Richards, 1939 |
| Lowland | Brazil | 442 | 205 | Grubb et al., 1963 |
| Lowland | British Guiana | 452 | 248 | Grubb et al., 1963 |
| Lowland | St. Kitts | 581 | 420 | Grubb et al., 1963 |
| Lowland | Ecuador | 581 | 290 | Grubb et al., 1963 |
| Equatorial Rainforest | Brazil | 594 | —— | Cain et al., 1956 |
| Lowland | Brazil | 678 | —— | Grubb et al., 1963 |
| Tall Evergreen | Thailand | 750 | —— | Ogawa et al., 1965a |
| Lowland | Ecuador | 883 | 474 | Grubb et al., 1963 |

The densities of trees in the Panamanian forests are similar to those of other tropical forests (table 4.4), with the exception of the Riverine type. The Riverine forest sampled by us in the Darien had a much lower density than other examples of this forest. The Cativo, the dominant tree in this forest, is a valuable lumber species, and the fewer and larger trees in our plot may represent the undisturbed condition for this forest.

FOREST WATER CONTENT

The concentration of water in the vegetation varied with location and compartment (table 4.5). An analysis of variance (table 4.6) showed that the differences between sites, compartments, and the site-compartment interaction were significant at the 0.01 level of probability. Premontane Wet forest and the Tropical Moist forest during the wet season contained the highest percentage of water, and Mangrove and dry season Tropical Moist forest the least. The analysis of variance suggests that the observed differences are related to the site and, therefore, to the amount of available water, as well as to the amount of foliage. Premontane Wet forests and Tropical Moist forests are exposed to the largest quantities of rain input, while Riverine forests probably have an adequate water supply in the soil available year-round. On the other hand, the amount of foliage compared to wood is greatest in Premontane Wet and Tropical Moist forests.

Table 4.5.   Percent water in vegetation samples arranged by compartment in five Panamanian forests.

| Compartment | | Tropical Moist | | Premontane Wet | Mangrove | Riverine |
| --- | --- | --- | --- | --- | --- | --- |
| | | Dry season | Wet season | | | |
| Overstory leaves | X | 63.21 | 69.91 | 56.34 | 65.45 | 67.41 |
| | SE | 3.15 | 1.47 | 1.66 | 0.74 | 2.70 |
| Overstory stems | X | 48.35 | 66.73 | 59.09 | 40.30 | 61.00 |
| | SE | 0.35 | 7.92 | 1.65 | 0.37 | 4.96 |
| Understory leaves | X | 64.28 | 73.75 | 76.43 | 70.42 | 74.69 |
| | SE | 2.48 | 1.26 | 2.51 | 0.43 | 2.05 |
| Understory stems | X | 54.23 | 65.30 | 71.89 | 53.02 | 61.26 |
| | SE | 0.30 | 1.71 | 1.91 | 1.08 | 1.55 |
| Litter | X | 20.89 | 72.37 | 61.18 | 21.77 | 58.27 |
| | SE | 0.51 | 0.56 | 1.02 | 3.34 | 1.73 |
| Roots | X | 60.28 | 61.80 | 57.44 | 78.81 | 52.02 |
| | SE | 1.67 | 2.90 | 1.58 | 7.54 | 3.28 |

Table 4.6.   An analysis of variance of moisture
content in Panamanian forests.

| Source | df | Mean Square |
|---|---|---|
| Total | 368 | |
| Site | 8 | 1054* |
| Compartment | 5 | 1635* |
| Site *X* compartment | 40 | 296* |
| Error | 315 | 44 |

* Asterisks indicate significance of "F" at the 0.01 level of probability.

#### THE VEGETATION BIOMASS OF MATURE FORESTS

Before we examine the details of the biomass structure, it will be helpful to compare the overall distribution of organic weight in the five forests (table 4.7). The Tropical Moist forest differs from the other forests mainly in having a greater understory biomass. The Premontane Wet forest is much like the Tropical Moist forest in weight of leaves and stems; it differs from other forests in the large biomass of epiphytes reflecting a different moisture environment. The Riverine forest has a much greater overstory stem and litter biomass than the other types; overstory stems were almost ten times greater than in the other forests. The Mangrove forest has a lower biomass in all categories except the roots and litter. Prop roots are characteristic of the Mangrove forest. When prop root biomass is added to overstory stems, the combined weight almost equals the overstory stems in the Premontane Wet and Tropical Moist forests. The underground mangrove biomass is essentially a fibrous root mass and for this reason the weight of mangrove roots was much greater than that in other forests. The mangrove prop roots catch floating debris brought into the forest by the tide; this debris resulted in the high litter biomass.

Each of the forest vegetation compartments can be examined in more detail. Understory biomass was approximately equal on each half of the one-quarter hectare plot (table 4.8) suggesting a uniformity of biomass over areas of this size. Greatest stem weight per hectare was encountered in the Tropical Moist forest, although stems varied between the two sites in this forest type. At the Rio Sabana site, the understory was largely dominatd by woody species like *Piper* spp.,

Table 4.7. Biomass of five Panamanian forests (kg/ha, dry weight). Data are presented as calculated; they do not represent accuracy to one kilogram per hectare.

| | Tropical Moist | | Premontane Wet | Riverine | Mangrove |
|---|---|---|---|---|---|
| Compartment | Rio Sabana | Rio Lara | | | |
| Overstory leaves | 7349 | 11367 | 10576 | 11383 | 3547 |
| Overstory stems | 252126 | 354735 | 258434 | 1163841 | 159294 |
| Understory leaves | 744 | 620 | 347 | 683 | 3 |
| Understory stems | 3271 | 1094 | 563 | 559 | 15 |
| Epiphytes | * | 2 | 1440 | 79 | 21 |
| Fruits and flowers | 8 | 139 | 7 | 30 | 21 |
| Prop roots | — | — | — | — | 116432 |
| Roots | 12633 | 9850 | 12707 | 12185 | 189761 |
| Total living biomass | 276131 | 377807 | 284074 | 1188760 | 469094 |
| Litter (excluding large dead wood) | 6200 | 2910 | 4820 | 14146 | 102106 |
| Large fallen dead wood | * | 14644 | * | 4929 | * |
| Total dead biomass | 6200 | 17554 | 4820 | 19075 | 102106 |

* Not determined.

Table 4.8. Understory biomass of five Panamanian forests (kg/ha dry weight). Replications indicate ⅛ hectare subplots used for understory sampling.

| | Tropical Moist | | Premontane Wet | Riverine | Mangrove |
|---|---|---|---|---|---|
| Component | Rio Sabana | Rio Lara | | | |
| Stems—Rep. 1 | 1510 | 634 | 305 | 289 | 8 |
| Stems—Rep. 2 | 1761 | 460 | 258 | 270 | 7 |
| Totals | 3271 | 1094 | 563 | 559 | 15 |
| Leaves—Rep. 1 | 321 | 354 | 197 | 384 | 1 |
| Leaves—Rep. 2 | 423 | 266 | 150 | 299 | 2 |
| Totals | 744 | 620 | 347 | 683 | 3 |
| Ratio leaves/stems | .23 | .57 | .62 | 1.22 | .19 |

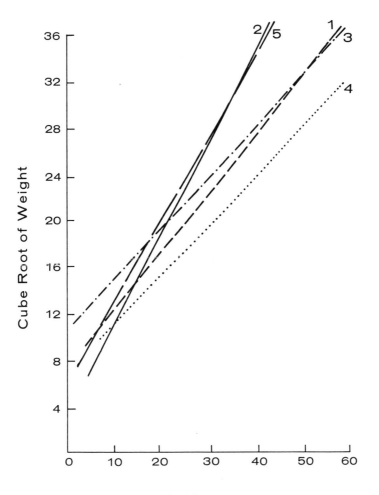

Figure 4.8. Relation of leaf biomass to stem diameter. Regression of the cube root of oven dry weight of leaf biomass on tree diameter (DBH) in five Panamanian forests. Number 1 refers to Tropical Moist forest at Rio Sabana site, 2 to Premontane Wet, 3 to Mangrove, 4 to Riverine and 5 to Tropical Moist forest at Rio Lara site.

while at the Rio Lara site fleshy plants were more abundant. Leaf biomass was greatest at the Tropical Moist forest Rio Sabana site. A great quantity of *Dieffenbachia Pittieri* also was encountered and, since all of the plant is green, it was included in the leaf weights. Without *Dieffenbachia*, the total leaf weight was 460 kg/ha. The ratio of leaves to stems varied widely between forests (table 4.8). In the Riverine forest, understory leaves had a greater weight than did stems. In all other locations, weight of stems exceeded weight of leaves. Mangrove forest had a very poorly developed understory and stem weight was five times leaf weight.

The biomass of the overstory vegetation was determined by regression of biomass on DBH and the frequency distribution of tree diameters as described earlier. The regression slope of leaf weight and tree DBH was similar in the Riverine and Mangrove forests and in the Premontane Wet and Rio Lara Tropical Moist forests (fig. 4.8). The Rio Sabana Tropical Moist forest was intermediate between these two groups. The Premontane Wet and Rio Lara Tropical Moist forest had the greatest weight of leaves for a given diameter.

Table 4.9.  Ratio of the leaf to stem biomass in the overstory
of forests in eastern Panama.

| Diameter class cm | Tropical Moist | | Premontane Wet | Riverine | Mangrove |
|---|---|---|---|---|---|
| | Dry season | Wet season | | | |
| 0-1 | —— | 0.33 | 0.41 | 0.37 | —— |
| 1-2 | —— | 0.21 | 0.34 | 0.30 | —— |
| 2-3 | 0.29 | —— | 0.23 | 0.26 | —— |
| 3-5 | 0.16 | 0.18 | 0.15 | 0.21 | 0.16 |
| 5-7 | 0.08 | 0.12 | 0.11 | 0.14 | —— |
| 7-10 | 0.11 | 0.15 | 0.09 | 0.11 | 0.06 |
| 10-15 | 0.12 | 0.12 | 0.06 | 0.07 | 0.11 |
| 15-20 | 0.05 | 0.09 | 0.02 | 0.03 | 0.07 |
| 20-30 | 0.02 | 0.04 | 0.07 | 0.03 | 0.02 |
| 30-40 | 0.03 | 0.05 | 0.08 | 0.04 | 0.03 |
| 40-60 | —— | 0 | 0 | 0.05 | —— |
| 60+ | —— | 0.02 | —— | 0.01 | —— |
| All diameters | 0.03 | 0.03 | 0.04 | 0.01 | 0.02 |

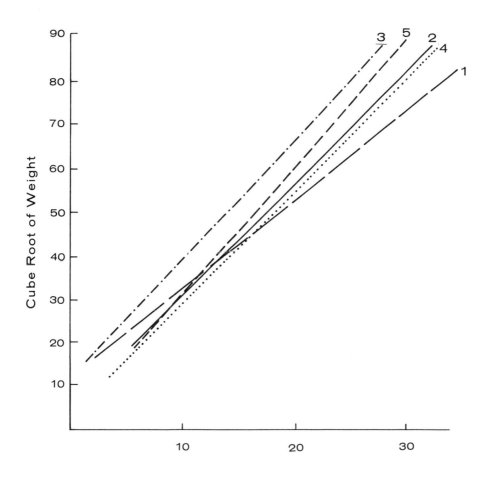

DBH in cm.

Figure 4.9.   Relation of stem biomass and tree diameter. Regression of the cube root of oven dry stem biomass on tree diamter (DBH) in five Panamanian forests. Number 1 represents Tropical Moist forest at Rio Sabana; 2, Premontane Wet; 3, Mangrove; 4, Riverine; and 5, Tropical Moist forest at Rio Lara.

The relationship between biomass of stems and diameter was similar for all the forests (fig. 4.9). Actually, the two Tropical Moist forests bracketed the regression coefficients for the other sites. The regression for stems was considerably greater than that for leaves, showing that biomass of stems per diameter increased at a greater rate than did the biomass of leaves in the forests.

In each forest, the ratio of overstory leaf biomass to the total stem and branch biomass tended to decrease with an increase in diameter of the tree, (table 4.9). For example, the ratio of leaves to stems from 0 to 3 centimeter diameter was about 0.30 and from 30 to 40 centimeters, about 0.05. Even so, the sites did not differ greatly. The ratios for the entire vegetation were all quite low and of the same order. Compared to the understory the ratios of leaves to stems in overstory were lower for most diameters and forests.

Vines were not harvested separately, but were combined with the overstory biomass. Ogawa et al. (1965b) have shown that while vines reduce the mass of leaves on the supportive tree, the vacant space is taken up by the liana's own leaves. For this reason, the combined leaf mass of trees and associated lianas were estimated from tree diameters.

In tropical forests, there may or may not be one single period during which all the plants flower and produce fruit (Janzen, 1967). For this reason, it is difficult to determine the average annual standing crop of these materials by the type of sampling employed here. Fruits and flowers were removed from the trees that were harvested at all sites. In order to calculate the biomass of fruits and flowers per area, it was necessary to convert the weights collected from harvested trees to those representing an area of forest. Our field observations suggest that a bimodal periodicity of flowering occurs in the Darien forests where the vegetation responds to a distinct dry season. According to Snow (1965) vegetation on Trinidad, transitional between Lower Montane rain forest and lowland seasonal forest, also has two seasons of tree flowering. Therefore, total production of fruits and flowers is perhaps about double the standing crop. Janzen (1967) stresses synchronization of sexual reproduction of trees during the dry season, which if applied to our situation will mean that fruit biomass is probably underestimated. If these estimates of fruits and flowers are approximately correct, the Tropical Moist forest wet-season site had a greater quantity of fruit and flowers (139 kg/ha) than did the Premontane Wet (7 kg/ha), the Riverine (30 kg/ha) or the Mangrove (21 kg/ha).

In this study, epiphytes included bromeliads, orchids, and similar plants; vines were not included. Epiphytes do not occur equally on all trees and we were unable to harvest all stems with epiphytes. Therefore, it was necessary to expand the total weights of epiphytes harvested per plot to the area of forest. We have assumed that we harvested a representative sample of trees and that epiphyte biomass is related to stem biomass since epiphytes depend on the stems for support. Ratio of the biomass of harvested epiphytes to the biomass of harvested stems was determined and multiplied by the total biomass of stems. With these assumptions, the greatest quantity of epiphytes was collected in the Premontane Wet forest (table 4.10) where the ratio of epiphytes to stems was about 5.6 percent. In this forest, the biomass of epiphytes far exceeded that of fruit and flowers. Lowest biomass of epiphytes was encountered in the Tropical Moist forest (table 4.10).

Table 4.10.    Biomass of epiphytes in four Panamanian forests.

| Forest type | Harvested weight of epiphytes (g wet wt.) | Ratio epiphytes to stems × 100 | Estimated biomass epiphytes (kg dry wt/ha) |
|---|---|---|---|
| Tropical Moist (wet season) | 587 | 0.0045 | 1.6 |
| Premontane Wet | 53118 | 5.5600 | 1440.0 |
| Riverine | 1591 | 0.0676 | 78.5 |
| Mangrove | 1337 | 0.1420 | 21.2 |

Standing crop of litter collected on the plots on the forest floor varied by a factor of 100 between sites (table 4.11). Greatest amounts of litter were collected in the Riverine and Mangrove forests. Both of these systems are periodically inundated by waters carrying debris, which could accumulate in the forest. This was especially true in the Mangrove forest where the litter included chairs, canoes, wood carvings, and other articles. Obviously logs and other natural debris also float into the forest. The litter in the Riverine forest was partly composed of heavy seeds of the dominant tree, *Prioria copaifera*. The actual quantity of these seeds was not determined at the time of harvest;

however, samples collected in May showed that 1.3 to 1.7 seeds fell per m$^2$ and that each seed weighed, on the average, 22 g dry weight.

In two sites, the Rio Lara Tropical Moist forest and the Riverine forest, all dead material over 2.5 cm diameter was collected and weighed in each replication. At the former site, this dead material weighed 1,464 g/m$^2$; and in the latter, 546 g/m$^2$. In the Rio Sabana Tropical Moist site, there was one large fallen tree on the area. It was estimated from measurements of the bole and limbs that this log weighed 8,165 kg wet weight. Assuming 50 percent moisture, the log contributed 408 g/m$^2$ to the standing crop of litter on the plot. There were no large dead fallen logs at the Premontane Wet or Mangrove sites.

Table 4.11.    Standing crop of litter (kg/ha) in five forests
in eastern Panama.

| | Tropical Moist | | Premontane Wet | Riverine | Mangrove |
|---|---|---|---|---|---|
| | Dry season | Wet season | | | |
| Number of samples | 10 | 10 | 10 | 10 | 20 |
| $\overline{X}$ | 6200 | 2910 | 4820 | 14150 | 102110 |
| SE | 1240 | 390 | 1050 | 1680 | —— |
| CV% | 63 | 42 | 69 | 37 | 96 |

The root study was designed to determine the biomass of surface roots; deep supporting roots were not measured and therefore the root biomass is underestimated. Biomass of roots 0 to 0.3 m varied between sites (table 4.12), with maximum root biomass in Mangrove forests. Undoubtedly some of the material classed as mangrove roots was dead roots and peat. *Rhizophora* has an extensive capillary root system which forms a thick fibrous peat-like soil (Rosevear, 1947). Where the peat stays moist, the decomposition of this fibrous peat is very slow and the organic carbon level and the content of fibrous roots in the peat remains high (Giglioli and Thornton, 1965). In the first 15 cm of the peat the weight of mangrove root material was approximately twice that in the next 15 cm. There was no statistical difference in total root biomass among the other four sites.

The Mangrove forest also is characterized by a highly developed prop root system which supports the trees. The prop roots were

Table 4.12.   Root biomass in five Panamanian forests (kg/ha dry wt.).

|  | Tropical Moist | | Premontane Wet | Riverine | Mangrove |
|---|---|---|---|---|---|
|  | Rio Sabana | Rio Lara | | | |
| Number of samples | 10 | 10 | 10 | 10 | 20 |
| $\overline{X}$ | 12633 | 9850 | 12707 | 12185 | 189761 |
| cv (%) | 58 | 48 | 73 | 31 | 31 |

measured separately, since they formed a structure unique to this forest. Prop roots weighed 116,432 kg/ha. These prop roots surpassed in weight the underground root biomass in the Mangrove forest.

Comparison of the biomass in these tropical forests with that of other forest communities (table 4.13) shows that leaf biomass of Tropical Moist (wet season), Premontane Wet, Riverine, and Temperate Coniferous forest, while similar to each other, were greater than that of other forests. Stem biomass was larger in the Riverine forest than in any other type. Most of the tropical forest examples, and all of those reported in this study, had a stem biomass larger than that of the three temperate forest types. Understory biomass was greatest in Tropical Moist forest and in Pine forest, but in each case the understory biomass was considerably less than the overstory leaf biomass. With the exception of the Mangrove, the root biomass of the Panamanian forests was about one-half that reported for other tropical and temperate forests; however, the roots have been shown to be underestimated. Again, with the exception of Mangrove, litter biomass in the tropical forests was a factor of two to ten below that in temperate forests, reflecting the more rapid turnover of litter under tropical conditions.

Comparison of the tropical forest biomass with that of other biomes in a transect from Arctic to Equator (table 4.14) illustrates the world pattern of biomass distribution. There is a progressive increase in vegetation biomass from tundra to tropical forest; the averages calculated from data in Rodin and Bazilevich (1967) show the range between extremes is about a factor of 20. Note, however, that the average biomass of tropical forests by these authors is greater than the average (342 wt/ha) of forests in table 4.13. The percentage of this

Table 4.13.    Biomass of forest communities. The data are presented as metric tons dry wt/ha (one metric ton equals 1000 kg).

| Forest | Canopy Leaves | Canopy Stems | Understory | Roots | Litter | Reference |
|---|---|---|---|---|---|---|
| **Tropical forests** | | | | | | |
| Thailand, Rainforest | 8.2 | 360 | 2.4 | 33 | 3.5 | Ogawa et al. (1965b) |
| Thailand, Monsoon forest | 3.8 | 261 | 2.0 | 25 | —— | Ogawa et al. (1965b) |
| Thailand, Dry Evergreen forest | 5.6 | 229 | 2.9 | —— | —— | Sabhasri et al. (1968) |
| Puerto Rico, Mangrove | 5.4 | 40 | —— | 4 | —— | Golley et al. (1962) |
| Ghana, Moist Tropical forest | —— | 187 | —— | 25 | 2.3 | Greenland and Kowal (1960) |
| Congo, Secondary forest | 6.5 | 116 | —— | 31 | 5.6 | Greenland and Kowal (1960) |
| Puerto Rico, Lower Montane forest | 8.1 | 269 | —— | 71 | 19.3 | Odum et al. (1970) |
| Brazil, Mountain Evergreen forest | 9.1 | 131 | —— | 33 | —— | Rodin and Bazilevich (1967) |
| **Panama** | | | | | | |
| Tropical Moist dry | 7.3 | 252 | 3.9 | 13 | 6.2 | This study |
| Tropical Moist wet | 11.3 | 355 | 1.7 | 10 | 2.9 | This study |
| Premontane Wet | 10.5 | 258 | 0.8 | 13 | 4.8 | This study |
| Riverine | 11.3 | 1163 | 1.1 | 12 | 14.1 | This study |
| Mangrove | 3.5 | 159 | .1 | 190 | 102.1 | This study |
| **Temperate forests** | | | | | | |
| Pine forest | 6.4 | 122 | 4.0 | 29 | 37.5 | Ovington (1965) |
| Coniferous forest | 10.4 | 114 | 2.0 | 38 | 36.6 | Ovington (1965) |
| Deciduous forest | 2.6 | 142 | 3.0 | 37 | 10.5 | Ovington (1965) |

biomass in green parts decreases from tundra to temperate deciduous forest but appears to increase again in the tropical forest. These differences represent adaptations to environment and the conditions of production at different latitudes. They suggest that temperate forests support the largest amount of biomass for the smallest biomass of leaves.

## BIOMASS OF SECOND GROWTH TROPICAL MOIST FOREST*

Throughout the tropical world, forested lands are cultivated by a system of shifting agriculture in which clearings made in the forest are cropped for a shorter number of years than they are fallowed (Conklin, 1963). According to Guzman (1956), *cultivo de roza,* shifting, or slash-burn agriculture, is the basic farming practice in Panama. In the Darien and San Blas Provinces, all agriculture except that adjacent to the few permanent settlements is the shifting type. The major crops are maize, rice, plantains, and bananas. Since only about two percent of the land area is in agriculture or second-growth forest resulting from recent farming, second-growth forest is not very significant today in terms of land area. However, it will probably become increasingly important as more of the province is settled.

Most second-growth vegetation in eastern Panama and northwestern Colombia occurs below an elevation of 100 meters, since cultivated fields are normally located on level terrain adjacent to the coasts and navigable rivers. In more populated areas, farmlands extend from the water's edge onto moderately to steeply sloping hillsides. The vegetation on a site to be planted is cut near the onset of the dry season in January, allowed to dry, and is burned during March or April. One or two crops are grown the first year and sometimes another the second year. Regrowth of natural vegetation often begins with the first crop. Fields may be allowed to remain fallow for six or more years.

Although there are numerous reports on shifting agriculture (see bibliographies of Conklin, 1963; Bartlett, 1955-61; Edwards and Rasmussen, 1942; and MacLeish et al., 1940), there are fewer studies on the natural fallow, or second-growth vegetation following agriculture (Bartholomew et al., 1953; Nye and Greenland, 1960; Tergas, 1965; Ewel, 1971). It is important in the comparison of tropical forests, to understand the dynamics of the succession stages leading to the stable Tropical Moist forest type.

Four second-growth sites were harvested in Panama during 1967. They included two sites where vegetation regrowth had occurred over two years, and two sites where regrowth had occurred over four and six years. The first two-year site and the four-year and six-year sites were harvested in July; the second two-year site was harvested in October. The sites were located in Tropical Moist forest within three kilometers of Santa Fe, Darien Province (latitude 8°39', longitude

* Prepared with John Ewel, University of Florida.

Table 4.14.    Vegetation biomass and percent of the biomass
in green parts in major vegetation regions.
Based on data from Rodin and Bazilevich (1967).

| Vegetation | Biomass wt/ha | % Biomass in green parts |
|---|---|---|
| Tundra | 28 | 11 |
| Northern coniferous forest | 137 | 6 |
| Temperate deciduous forest | 224 | 2 |
| Tropical wet forest | 564 | 5 |

78° 09′). This study was a cooperative venture between the University of Florida and the University of Georgia and some of the data has been reported by Ewel (1971).

The general aspect of young (10 years old) second-growth vegetation in eastern Panama was a low, dense growth of small trees, vines, and herbaceous plants. The canopy at each site was dominated by *Cecropia spp.*, but contained other common genera including *Ochroma, Trema, Spondias*, and *Persea*. Palms such as *Bactris* were present but not common. *Heliconia* and *Calathea* were abundant at all sites, as were herbaceous and woody vines. Scattered large trees are often left standing after the initial clearing of the mature forest. Nonvascular stem epiphytes were common, but epiphyllae were almost nonexistent. All sites were on slightly undulating terrain which appeared to result from disintegration of large roots, stumps, and logs left after the original clearing, and from fallen trees when the area was in mature forest.

The first two-year site, harvested in July, was originally cleared of forest at least eight years prior to this study and had gone through several plantings. Charred stumps and large boles from the original forest were still present (fig. 4.10), as were maize stalks from the last planting. At the time of harvest a dense growth of vegetation with a low canopy reaching a height of five meters existed. The aspect of the vegetation was markedly uneven and dense stands of grass or low-growing herbs, under one meter in height, were common. At about seven meters, the taller trees formed an open canopy of scattered emergent crowns.

Figure 4.10. Two-year-old site after the understory harvest in July. Note the charred boles and stumps still remaining from the original clearing from forest. A large tree that was not cut during the original clearing may be seen in the background. (Photograph by Peter McGrath.)

The second two-year site, harvested in October, had been planted only once since it was cleared. Stumps and fallen boles of forest trees were still much in evidence. The general aspect of the vegetation on this site was that of a better developed forest than the first two-year site. The ground flora was relatively open, with the bulk of the leaves and vines located in the canopy, approximately eight meters (range 3-10 meters) in height.

The four-year site, cropped once since clearing, was generally similar to the second two-year site, except that it was structurally more uniform. Canopy height averaged about ten meters (range 5-12 meters), or almost the same as Ross (1954) reported for five-year-old second-growth in Nigeria.

The six-year fallow, which has been planted twice, was the oldest second-growth stand harvested. The understory was more open than that of younger stands (fig. 4.11) and the average height of the relatively even canopy was about 12 meters (range 10-13.5 meters) (fig. 4.12). Several palms *(Scheelea sp.)*, which had not been killed during the original clearing, had grown to a considerable size. Species of plants observed in these second growth stands are listed in appendix table 5.

Table 4.15.    Structural features of second-growth vegetation
in eastern Panama.

| Age of stand in years | 2 | 2 | 4 | 6 |
|---|---|---|---|---|
| Month of harvest | July | October | July | July |
| Leaf area index ($m^2/m^2$) | 7.5 | 6.9 | 11.6 | 16.5 |
| Percent cover | 66 | 80 | 80 | 84 |
| Crown diameter of largest trees (m) | 2 | 4 | 4 | 6 |
| Height of largest trees (m) | 7.2 | 10.1 | 12.3 | 13.6 |
| DBH of largest trees (cm) | 6 | 12 | 13.5 | 17.5 |

The structural characteristics measured in the Panama second-growth (table 4.15) changed rapidly with age of the vegetation. The leaf area index of four- and six-year-old stands (11.5 and 16.5

Figure 4.11.  Second-growth stand being harvested near Santa Fe, Darien, Panama. (Photograph by the authors.)

Fig. 4.12  Six year old second growth stand after harvest of the plot. The large trees in the background were not cut when the original agricultural field was established.

$m^2/m^2$) were nearly the same as the leaf areas observed in mature forests. In contrast, height, and DBH were three to four times lower than that observed in Tropical Moist forests. These data suggest that the production capacity of the forest is quickly reestablished while the supporting structure of stems and branches is more slowly developed.

Table 4.16.   Dry-weight biomass (kg/ha) by compartment and age of second-growth in eastern Panama.

| Age of stand | 2 | 2 | 4 | 6 |
|---|---|---|---|---|
| Month of harvest | July | October | July | July |
| Overstory | | | | |
| Leaves | 1200 | 2200 | 3300 | 4800 |
| Stems | 4300 | 18000 | 22700 | 27100 |
| Fruits and flowers | 10 | 40 | 30 | 140 |
| Understory | | | | |
| Leaves | 2400 | 800 | 2600 | 1700 |
| Stems | 5100 | 3300 | 9300 | 8800 |
| Fruits and flowers | 10 | 20 | 110 | 10 |
| Roots | 2600 | 4100 | 4500 | 14200 |
| Total live biomass | 15620 | 28460 | 42540 | 56750 |
| Litter | 4600 | 4400 | 5700 | 6100 |

The results of the standing crop determinations are shown in table 4.16. Biomasses of compartments and sites were significantly different. With two exceptions, all compartments and stand totals increased with age. The six-year field, although highest in total biomass, was lower than the four-year stand in all understory compartments. The second deviation from the general trend was the low weight (800 kg/ha) of understory leaves in the two-year stand harvested in October. Both replications at this site had low biomass of understory leaves: 750 and 850 kg. Although some of the variation between the two-year-old stands may have resulted from growth from July to October, it also probably reflects the variation in site quality. The October site was located in a large area of two-year fallow, portions of which appeared similar to the July two-year harvest site, while other parts of the same field resembled the four-year stand.

Much of the increase in total biomass from four to six years is accounted for by the high biomass of roots in the six-year fallow field. Above ground biomass increased only 4,500 kg/ha from four to six years, as opposed to an increase of 14,200 kg/ha for total biomass.

The biomass values for fruits and flowers (20 to 150 kg/ha) were considerably greater than observed for most of the mature forests. However, Tropical Moist forest harvested at about the same time during the wet season had a fruit and flower biomass of 139 kg/ha. Fruit and flowers in second growth were mainly from *Heliconia*, and in the six-year stand, from the palm *Scheelea*. Other genera probably accounted for less than five percent by weight of the inflorescences.

Table 4.17.    Dry weight standing crop of second-growth vegetation in Panama and the Congo (Bartholomew et al., 1953).

| | | Standing crop (kg/ha) | | | |
|---|---|---|---|---|---|
| Age | Location | Leaves | Stems | Roots | Litter |
| 2 | Panama-July | 3600 | 9400 | 2600 | 4600 |
| 2 | Panama-October | 3000 | 21300 | 4100 | 4400 |
| 2 | Congo | 5560 | 12299* | | ―― |
| 4 | Panama | 5900 | 32000 | 4500 | 5700 |
| 5 | Congo | 5627 | 71067 | 25753 | 7320 |
| 6 | Panama | 6500 | 35900 | 14200 | 6100 |
| 8 | Congo | 5379 | 116313 | 22682 | 7983 |

* Representing both stems and roots.

The standing crops of second-growth Tropical Moist forest vegetation can be compared with similar data from the Congo (Bartholomew et al., 1953) (table 4.17). In general, the values for successional stands in the Congo basin do not differ greatly from those we obtained in Panama. However, the standing crop of stems of five and eight-year Congo vegetation was two to four times that of six-year vegetation in Panama. This was also true for six- and seven-year-old stands studied by Kellman (1970) in Mindanao.

# CHAPTER 11

# Forest Chemical Comparisons

In section 2 we observed that concentrations of the elements differed by compartment within the Tropical Moist forest. Here we will expand this comparison and examine the concentrations in each of the types of forests studied in Darien Province. An analysis of variance of concentration over site and compartment showed that site and compartment both accounted for a highly significant ($P < 0.01$) amount of the variation for all elements, except nickel and titanium. The effects of site and compartment were not independent of each other since the site-compartment interaction also was found to be highly significant ($P < 0.01$).

Considering first the differences between sites, second growth and Tropical Moist forest had highest total concentration of elements, and Premontane Wet forest the lowest, in the organic compartments (table 4.18). These differences were mainly due to the higher concentrations of calcium and potassium occuring in Tropical Moist and second-growth forests, reflecting the nature of the soil in this area. In contrast, Premontane Wet forest had larger concentrations of sodium, aluminum, iron, and manganese than the other terrestrial forests. Mangrove forests differed greatly from the other forests, having lower concentrations of phosphorus, potassium, and calcium and higher concentrations of magnesium, sodium, boron, iron, and strontium. Copper, nickel, lead, titanium, and zinc exhibited fewer differences between sites than the other elements. The most concentrated elements, considering the overall means for all sites, were calcium, potassium, magnesium, phosphorus, sodium, aluminum, iron, and manganese (table 4.18).

Analyses for nitrogen were incomplete since they were made at the end of the study and sufficient material for all sites and compartments was no longer available. The analyses by site are presented in table 4.19. The overall mean concentration was 1.4 percent nitrogen. The only site which deviated greatly from this mean was Mangrove forest which had 0.7 percent nitrogen in leaves, stems, fruits and flowers,

Table 4.18.   Concentration of elements in forests.
(ppm dry weight) in the organic compartments
In each column forests sharing the same letter are not

| Forest | P | K | Ca | Mg | Na | Al | B | Ba |
|---|---|---|---|---|---|---|---|---|
| Tropical Moist Rio Sabana | 1200 $_c$ | 10700 $_c$ | 19000 $_a$ | 2000 $_c$ | 200 $_d$ | 1059 $_b$ | 27 $_b$ | 37 $_e$ |
| Tropical Moist Rio Lara | 1200 $_c$ | 11400 $_c$ | 16600 $_b$ | 1900 $_c$ | 200 $_d$ | 1040 $_b$ | 17 $_{cd}$ | 96 $_c$ |
| Riverine | 1600 $_b$ | 11200 $_c$ | 15400 $_{bc}$ | 2300 $_{bc}$ | 600 $_c$ | 993 $_b$ | 18 $_{cd}$ | 62 $_d$ |
| Premontane Wet | 1000 $_{cd}$ | 8300 $_{cd}$ | 10500 $_d$ | 2700 $_b$ | 1800 $_b$ | 1549 $_a$ | 23 $_{bc}$ | 113 $_b$ |
| Mangrove | 900 $_d$ | 5400 $_d$ | 9400 $_d$ | 3500 $_a$ | 8300 $_a$ | 930 $_b$ | 48 $_a$ | 20 $_f$ |
| Two-year second growth, July | 2000 $_a$ | 19400 $_a$ | 14000 $_{bc}$ | 2700 $_b$ | 100 $_d$ | 567 $_c$ | 20 $_{cd}$ | 89 $_c$ |
| Two-year second growth, October | 1700 $_b$ | 14200 $_b$ | 13800 $_c$ | 2600 $_b$ | 200 $_d$ | 604 $_c$ | 17 $_{cd}$ | 136 $_a$ |
| Four-year second growth | 2100 $_a$ | 19300 $_a$ | 13200 $_c$ | 2300 $_{bc}$ | 100 $_d$ | 399 $_c$ | 15 $_d$ | 96 $_c$ |
| Six-year second growth | 1600 $_b$ | 14900 $_b$ | 14400 $_{bc}$ | 2300 $_{bc}$ | 200 $_d$ | 337 $_c$ | 21 $_{bcd}$ | 124 $_{ab}$ |
| Overall mean | 1400 | 12900 | 14200 | 2500 | 1000 | 822 | 22 | 87 |

* Not measured.

and roots. The overall mean for nitrogen was about the same reported
by Rodin and Bazilevich (1967), 1.22 percent, for tropical vegetation.

MEAN COMPARTMENTAL CONCENTRATIONS OF THE ELEMENTS

The mean compartmental concentrations are given in table 4.20.
Understory fruits and flowers and understory leaves have the highest
concentrations, while overstory stems have the lowest concentrations.
On the basis of the data in table 4.20 the compartments can be ranked
according to highest element content as follows: understory fruits
and flowers > understory leaves > overstory leaves > litter > overstory
fruit and flowers > roots > understory stems > overstory stems. Understory fruit and flowers had highest concentrations of calcium, aluminum, iron, molybdenum, and strontium, while overstory leaves had
highest concentrations of boron and cesium. Roots had highest concentrations of sodium and zinc. Cobalt, nickel, lead, and titanium
were distributed relatively evenly over the compartments.

Comparison of the average elemental concentration
in different tropical forests in Darien, Panama.
statistically different for that element at the 5 percent level.

| Co | Cs | Cu | Fe | Mn | Mo | Ni | Pb | Sr | Ti | Zn | Total measured ppm |
|---|---|---|---|---|---|---|---|---|---|---|---|
| 31 e | —* | 9 bc | 218 cd | 56 d | 4.5 b | 25 a | —* | 103 b | 13 a | 28 bc | 34711 |
| 51 c | 18 d | 7 c | 133 d | 71 d | 4.0 bc | 7 b | 36 bc | 78 c | 6 a | 26 c | 32890 |
| 71 a | 24 bc | 12 ab | 390 b | 68 d | 3.6 bc | 78 a | 38 abc | 88 bc | 19 a | 36 abc | 33001 |
| 50 c | 30 a | 10 bc | 511 a | 352 a | 6.9 a | 23 a | 44 ab | 136 a | 25 a | 43 a | 27216 |
| 61 b | 22 c | 15 a | 464 ab | 209 b | 4.3 bc | 7 b | 60 a | 146 a | 39 a | 40 ab | 29565 |
| 48 c | 33 a | 12 ab | 165 cd | 187 bc | 2.0 ce | 8 b | 24 c | 73 c | 9 a | 30 bc | 39467 |
| 39 d | 27 b | 13 a | 263 c | 165 bc | 2.9 bcd | 9 b | 25 c | 101 b | 10 a | 30 bc | 33942 |
| 61 b | 25 bc | 14 a | 197 cd | 195 bc | 2.5 cd | 29 a | 56 a | 51 d | 10 a | 35 abc | 38186 |
| 31 e | 22 c | 12 ab | 137 d | 154 c | 1.5 d | 7 b | 30 bc | 105 b | 126 a | 29 bc | 34543 |
| 49 | 25 | 12 | 267 | 159 | 3.6 | 49 | 39 | 97 | 29 | 33 | 33694 |

Table 4.19.  Nitrogen concentration in the biotic compartments
of forests in Darien, Panama.

| Site | Number of samples | Mean percent nitrogen |
|---|---|---|
| Tropical Moist (Rio Lara) | 66 | 1.4 |
| Premontane Wet | 60 | 1.5 |
| Mangrove | 45 | 0.7 |
| Second growth, 2-year-old, harvested July | 46 | 1.6 |
| Second growth, 2-year-old, harvested October | 42 | 1.6 |
| Second growth, 6-year-old | 62 | 1.5 |
| Overall mean | | 1.4 |

Table 4.20.   Comparison of mean concentration
of tropical forests in Darien, Panama. Compartments
statistically different

| Compartment | P | K | Ca | Mg | Na | Al | B | Ba |
|---|---|---|---|---|---|---|---|---|
| Overstory leaves | 1600 c | 13500 c | 16600 b | 3100 b | 1600 b | 542 e | 44 a | 79 d |
| Understory leaves | 1600 c | 17000 b | 17000 b | 3300 ab | 1200 bc | 1194 c | 27 b | 100 bc |
| Overstory stems | 1200 d | 9000 e | 9600 c | 1100 d | 900 cd | 255 f | 7 e | 55 e |
| Understory stems | 1200 d | 11800 d | 10100 c | 1400 d | 1200 bcd | 864 d | 11 d | 87 cd |
| Overstory fruit and flowers | 2200 b | 16900 b | 8700 c | 2600 c | 900 de | 271 f | 16 c | 56 e |
| Understory fruit and flowers | 2600 a | 35800 a | 14300 b | 3800 a | 100 e | 144 f | 18 c | 130 ab |
| Litter | 1200 d | 3700 g | 24000 a | 2200 cd | 200 e | 1740 a | 26 b | 151 a |
| Roots | 1100 e | 6600 f | 13600 bc | 2600 c | 1700 a | 1302 b | 20 c | 71 cd |
| Overall mean | 1400 | 12900 | 14200 | 2500 | 1000 | 822 | 22 | 87 |

* Not measured.

Nitrogen concentration is reported separately because the method of analysis was different from the other elements and not all sites are represented. Leaves contained greater concentrations of nitrogen than the other compartments (table 4.21). Litter, fruit, and flowers were similar to leaves, while nitrogen in stems and roots was considerably below the foliage concentrations.

MEAN ELEMENTAL CONCENTRATION IN SOIL

The absolute value for the concentration of an element in the soil depends upon the extraction procedure. As mentioned earlier, in agricultural applications a neutral ammonium acetate extraction is often employed and the concentrations are interpreted as the quantities of an element that may be exchanged between soil and plant. The exchangeable quantity is less than the total concentration since many elements may be chemically bound in such a way as not to be fully extracted by ammonium acetate. In this study we were interested in

of elements in the vegetation components
sharing the same letter are not
at the 5 percent level.

| Co | Cs | Cu | Fe | Mn | Mo | Ni | Pb | Sr | Ti | Zn | Total measured ppm |
|---|---|---|---|---|---|---|---|---|---|---|---|
| 47ab | 34a | 9d | 120d | 163d | 3.0bcd | 70a | 39a | 110b | 85a | 24f | 37769 |
| 51ab | 24bc | 9d | 308c | 205c | 4.1b | 51a | 42a | 106b | 16a | 34d | 42271 |
| 48ab | 24bc | 7e | 41f | 77f | 1.4d | 66a | 32a | 72c | 43a | 20g | 22548 |
| 46ab | 26b | 10d | 209d | 130d | 2.3bcd | 47a | 31a | 72c | 120a | 29de | 27384 |
| ——* | ——* | 13c | 75e | 93e | 4.2b | 53a | ——* | 61e | 50a | 26ef | 32018 |
| —* | ——* | 23a | 102d | 384a | 1.8cd | 40a | ——* | 99b | 62a | 50b | 57654 |
| 54a | 22c | 13c | 755a | 269b | 8.0a | 26a | 43a | 156a | 31a | 39c | 34633 |
| 49ab | 18d | 18b | 532b | 77g | 3.9bc | 23a | 47a | 99b | 33a | 51a | 27944 |
| 49 | 25 | 12 | 267 | 159 | 3.6 | 49 | 39 | 97 | 29 | 33 | 33694 |

Table 4.21. Concentration of nitrogen in tropical forest vegetation. The means represent the data from six sites listed in table 4.19.

| Compartment | Number of samples | Mean percent nitrogen |
|---|---|---|
| Overstory leaves | 50 | 2.0 |
| Understory leaves | 54 | 2.1 |
| Overstory stems | 40 | 0.5 |
| Understory stems | 36 | 0.7 |
| Overstory fruit and flowers | 36 | 1.4 |
| Understory fruit and flowers | 23 | 1.5 |
| Litter | 39 | 1.7 |
| Roots | 43 | 0.8 |
| Overall mean | —— | 1.4 |

Table 4.22.    Chemical concentration (ppm) of the soil (0-30 cm) compartment. Forests sharing the same letter are not statistically different at the 5 percent level.

| Forest type | P | K | Ca | Mg | Na | Co | Cs | Cu | Fe | Mn | Pb | Sr | Zn |
|---|---|---|---|---|---|---|---|---|---|---|---|---|---|
| 2-year second growth, July | $7.6_b$ | $62_c$ | $6457_a$ | $718_a$ | $259_d$ | $3.0_b$ | $1.1_c$ | $41_a$ | $19_c$ | $52_{cd}$ | $1.1_b$ | $11_d$ | $4.0_{de}$ |
| 2-year second growth, October | $5.6_d$ | $48_c$ | $3848_b$ | $525_c$ | $38_f$ | $2.2_{cde}$ | $1.8_b$ | $4_b$ | $12_{cd}$ | $48_{cd}$ | $0.9_b$ | $14_c$ | $12.2_{cd}$ |
| 4-year second growth | $7.4_b$ | $46_b$ | $3787_b$ | $434_d$ | $217_{de}$ | $2.2_{cd}$ | $0.9_c$ | $2_b$ | $12_{cd}$ | $103_b$ | $1.1_b$ | $9_d$ | $93.0_a$ |
| 6-year second growth | $6.9_c$ | $86_{bc}$ | $809_d$ | $410_d$ | $184_e$ | $1.7_{def}$ | $1.7_b$ | $4_b$ | $11_{cd}$ | $42_d$ | $1.6_b$ | $11_d$ | $5.3_{cde}$ |
| Tropical Moist, Rio Sabana | $2.6_f$ | $41_c$ | $5854_a$ | $521_c$ | $462_b$ | $1.5_{ef}$ | * | $1_b$ | $6_d$ | $6_e$ | * | $20_b$ | $9.6_{cde}$ |
| Tropical Moist Rio Lara | $7.6_b$ | $17_a$ | $4232_b$ | $645_b$ | $50_f$ | $1.6_{ef}$ | $1.6_b$ | $2_b$ | $6_d$ | $8_e$ | $2.2_b$ | $16_c$ | $51.0_b$ |
| Riverine | $3.6_e$ | $76_{bc}$ | $1972_c$ | $574_c$ | $673_a$ | $2.2_{cde}$ | $1.0_d$ | $3_b$ | $5_d$ | $36_d$ | $6.8_a$ | $10_d$ | $1.6_e$ |
| Premontane Wet | $0.2_g$ | $69_c$ | $466_d$ | $302_e$ | $104_f$ | $2.6_{bc}$ | $0.8_c$ | $5_b$ | $58_b$ | $176_a$ | $0.6_b$ | $2_e$ | $1.9_e$ |
| Mangrove | $24.8_a$ | $17_c$ | $1947_c$ | $42_f$ | $332_c$ | $5.5_a$ | $18.8_a$ | $4_b$ | $102_a$ | $61_c$ | $6.4_a$ | $28_a$ | $12.8_c$ |

* Not determined.

Table 4.23.    Concentration (ppm) of chemical Values in a column sharing a letter are not significantly

| Components | P | K | Ca | Mg | Na | Al | B |
|---|---|---|---|---|---|---|---|
| Overstory leaves | $290_{bc}$ | $10200_b$ | $11000_{bc}$ | $3700_{ab}$ | $2800_{ab}$ | $1165_c$ | $47_a$ |
| Understory leaves | $950_a$ | $10100_b$ | $9800_{cd}$ | $4100_a$ | $3200_a$ | $1935_{ab}$ | $26_b$ |
| Overstory stems | $80_c$ | $5800_c$ | $5900_{de}$ | $1100_f$ | $1400_c$ | $1116_{cd}$ | $8_c$ |
| Understory stems | $360_b$ | $9500_b$ | $7300_{de}$ | $1600_{ef}$ | $2400_b$ | $1483_{bc}$ | $10_c$ |
| Overstory fruits and flowers | $930_a$ | $14800_a$ | $5500_c$ | $3200_{abc}$ | $900_{cd}$ | $501_d$ | $18_{bc}$ |
| Litter | $160_c$ | $5000_c$ | $14700_a$ | $2300_{cef}$ | $300_d$ | $2000_a$ | $25_b$ |
| Roots | $60_c$ | $5200_c$ | $13300_{ab}$ | $2600_{bce}$ | $1100_c$ | $1740_{abc}$ | $10_c$ |

the total quantities of the mineral materials to 30 centimeters depth in the forest and our acid extraction procedure gave *total* rather than exchangeable concentrations.

The test for differences between the concentrations in the soil at the various sites revealed a rather complicated pattern (table 4.22). Mangrove mud contained highest quantities of phosphorus, cobalt, cesium, iron, and strontium. Two-year-old second-growth soil had highest concentrations of magnesium, and copper, and with Tropical Moist forest, highest concentration of calcium; Tropical Moist forest, during the wet season, had highest concentrations of potassium; Riverine forest had highest sodium; and Premontane Wet forest, highest manganese. Note that the four second-growth soils and the two Tropical Moist forest soils, which all came from the same area and showed considerable similarity in the vegetation, were as different from each other as from the other forests. Other studies we have carried out in the Southeastern United States suggest that there may be considerable local heterogeneity in soil concentrations. If this is true in Darien Province, Panama, the number of soil samples may have been too few to adequately represent the mean condition. If this is the case, it follows that the soil concentration and vegetation concentration may not appear to be related. A relationship between a maximum concentration in vegetation and soil in the same forest type was observed only for iron and strontium (compare tables 4.18 and 4.22).

elements in Premontane Wet forest.
different at the 0.05 level of significance.

| Ba | Co | Cs | Cu | Fe | Mn | Mo | Pb | Sr | Ti | Zn |
|---|---|---|---|---|---|---|---|---|---|---|
| 81 bc | 80 a | 39 ab | 7 cd | 259 b | 371 b | 5.9 ab | 50 b | 152 abc | 14 c | 48 ab |
| 81 bc | 51 a | 30 ab | 8 cd | 314 b | 543 a | 6.9 ab | 83 a | 136 abc | 17 c | 52 a |
| 82 bc | 49 a | 27 bc | 5 d | 71 b | 308 bc | 3.1 b | 30 cd | 62 c | 4 c | 40 ab |
| 107 b | 36 a | 29 abc | 9 bc | 220 b | 304 bc | 3.3 b | 26 d | 104 bc | 11 c | 49 ab |
| 41 c | 31 a | 4 d | 15 a | 94 b | 269 bc | 10.0 ab | 26 d | 85 bc | 4 c | 27 b |
| 181 a | 39 a | 46 a | 13 a | 1079 a | 369 b | 11.4 a | 36 cd | 173 a | 61 a | 39 ab |
| 150 a | 44 a | 11 cd | 11 ab | 986 a | 172 c | 4.1 b | 41 bc | 161 ab | 32 b | 33 ab |

Table 4.24.   Concentration (ppm) of
Values in columns sharing the same litter

| Compartment | P | K | Ca | Mg | Na | Al | B |
|---|---|---|---|---|---|---|---|
| Overstory leaves | 1400 b | 12200 bc | 18400 a | 3200 a | 300 b | 641 c | 41 a |
| Understory leaves | 1400 b | 16400 ab | 20200 a | 3400 a | 1200 a | 2000 a | 22 b |
| Overstory stems | 1000 c | 8700 cd | 11000 b | 1100 c | 400 b | 166 c | 8 c |
| Understory stems | 1500 b | 8700 cd | 12100 b | 1600 c | 300 b | 1205 b | 10 bc |
| Overstory fruit and flowers | 1800 a | 21500 a | 9900 b | 2700 ab | 200 b | 148 c | 14 bc |
| Understory fruit and flowers | 1900 a | 27300 a | 9800 b | 4300 a | 400 b | ——* | ——* |
| Litter | 900 c | 1800 e | 11000 b | 1500 c | 100 b | ——* | ——* |
| Roots | 1100 c | 6200 d | 18000 a | 2000 bc | 1000 ab | 1392 b | 10 bc |

* Not measured.

MEAN ELEMENTAL CONCENTRATIONS WITHIN FORESTS

Since concentration has been shown to vary with forest and compartment (table 4.18 and 4.20), it is necessary to consider the distribution of elements among compartments within each forest. Data for Tropical Moist forest were presented earlier (section 2); here we will examine the chemistry of other Darien forests. In all of these comparisons nickel is not included because determinations were not available for all compartments at all sites.

*Premontane Wet Forest.* In the Premontane Wet forest the elements occurring at highest concentrations were potassium, calcium, magnesium, sodium, and aluminum. Levels of iron and manganese were comparatively high, while phosphorus concentrations were low (table 4.23). Stems usually contained lowest concentrations, while foliage, fruit, or litter and roots contained the highest concentrations of the elements. For example, potassium, phosphorus, magnesium, sodium, boron, manganese, and lead occurred at highest concentrations in foliage or fruit. Highest concentrations of calcium, aluminum, barium, copper, iron, strontium, and titanium were observed in litter or roots. Since titanium characteristically occurs at relatively high

chemical elements in Riverine forest.
are not significantly different (P = 0.05).

| Ba | Co | Cs | Cu | Fe | Mn | Mo | Pb | Sr | Ti | Zn |
|---|---|---|---|---|---|---|---|---|---|---|
| 67ab | 54cd | 46a | 9c | 188d | 53b | 3.0bc | 30c | 100a | 12c | 23a |
| 95a | 72b | 32b | 13bc | 918a | 144a | 7.4a | 40b | 125a | 41a | 57a |
| 31c | 91a | 14c | 10c | 75d | 18c | 1.3c | 28c | 61b | 6c | 26a |
| 65ab | 45d | 14c | 8c | 356c | 70b | 3.1bc | 24c | 64b | 13c | 40a |
| 40bc | 74b | 15c | 25a | 102d | 41bc | 1.5c | 27c | 54b | 7c | 38a |
| ——* | 69bc | 27b | 7c | 101d | 154a | ——* | 29c | 85b | ——* | 27a |
| ——* | 91a | 1d | 5c | 1069a | 127a | ——* | 68a | 149a | ——* | 12b |
| 63b | 73b | 1d | 18ab | 582b | 72b | 4.6b | 40b | 110a | 28b | 36a |

concentrations in soil (Bowen, 1966), the data suggest that the litter and root samples may have been contaminated by soil.

*Riverine Forest.* Potassium, calcium, magnesium, and phosphorus were present at highest concentrations in the Riverine forest (table 4.24). Aluminum and iron also were present at relatively high levels. Understory leaves, roots and overstory leaves had highest concentrations of almost every element, while elements in overstory stems (except for cobalt) were present in lowest concentrations.

*Mangrove Forest.* The Mangrove forest was characterized by high concentrations of sodium, calcium, and potassium (table 4.25). The root compartment contained significantly higher concentrations of nine of fourteen elements tested. Foliage also contained high concentrations of elements and in the case of magnesium, shared the highest position in the statistical comparison with the roots.

Rodin and Basilevich (1967), and Lamberti (1969) also have determined the concentrations of certain of these elements in leaves of Red Mangrove (*Rhizophora* sp.) in Cuba and Brazil (table 4.26). With the exception of phosphorus, calcium, and sodium, our data fall within the range found by these investigators. Concentrations of

Table 4.25.   Concentration (ppm) of
Values in columns sharing the same letter

| Compartment | P | K | Ca | Mg | Na | Al | B |
|---|---|---|---|---|---|---|---|
| Overstory leaves | 900 b | 8400 b | 12200 a | 4700 a | 9800 ab | 160 c | 70 b |
| Understory leaves | 800 b | 8500 b | 7800 b | 5000 a | 8310 b | 1371 b | 51 c |
| Overstory stems | 900 b | 3000 d | 12900 a | 1000 d | 5500 c | 40 c | 8 f |
| Understory stems | 1600 a | 3900 c | 5700 b | 2900 c | 9500 ab | 2000 a | 37 d |
| Overstory fruits and flowers | 700 b | 10100 a | 5900 b | 2900 c | 9600 ab | 141 c | 23 e |
| Understory fruits and flowers | 800 b | 7100 b | 6900 b | 3400 b | 9700 ab | ——* | ——* |
| Litter | 500 c | 11800 a | 4200 b | 3900 b | 19100 a | ——* | ——* |
| Roots | 700 b | 2300 d | 7500 b | 4600 a | 8500 b | 2000 a | 81 a |

* Not measured.

phosphorus, calcium, and sodium were lower in Panama. The greatest disparity was for phosphorus and sodium, which were both about one-half the concentrations observed by Lamberti (1969).

*Second-Growth Forests.* The overall concentration for several elements such as phosphorus and potassium in the four second-growth forests was greater than in other forest sites (table 4.18). However, a sequential increase in overall concentration with succession was not observed. Further, none of the elements individually showed a consistent increase or decrease from two to six years of second growth. For this reason, the four second-growth forests will not be examined separately.

Three sets of data are available for comparison with the second-growth concentrations from Darien. Tergas (1965) examined the concentration of macroelements in one-year-old second-growth at 10 sites near Lake Izabal, Guatemala. Snedaker and Gamble (1969) reported on concentrations in second-growth species collected in Colombia. Stark (1970) studied the chemistry of second growth on podzols in Surinam. Levels of phosphorus, potassium, copper, iron, strontium, and zinc are similar at the various locations (table 4.27) in Central America and Colombia. Concentrations of calcium, aluminum,

chemical elements in Mangrove forest.
are not significantly different (P = 0.05).

| Ba | Co | Cs | Cu | Fe | Mn | Mo | Pb | Sr | Ti | Zn |
|----|----|----|----|----|----|----|----|----|----|----|
| 1 b | 46 c | 24 b | 6 b | 82 c | 387 a | 1.6 b | 32 c | 175 c | 5 c | 11 b |
| 2 b | 56 c | 16 b | 6 b | 672 b | 125 d | 6.5 a | 27 c | 96 d | 30 bc | 15 b |
| 1 b | 52 c | 20 b | 7 b | 36 c | 168 c | 1.0 b | 39 c | 209 a | 4 c | 11 b |
| 2 b | 83 a | 26 a | 8 b | 1000 a | 255 b | 8.6 a | 47 b | 98 d | 51 b | 12 b |
| 1 b | 81 a | 32 a | 6 b | 82 c | 191 c | 1.0 b | 40 c | 90 d | 5 c | 11 b |
| ——* | 56 c | 16 b | 4 c | 45 d | 164 c | ——* | 27 c | 55 d | ——* | 7 b |
| ——* | 65 b | 19 b | 4 c | 666 b | 77 d | ——* | 53 b | 73 d | ——* | 12 b |
| 4 a | 61 b | 22 b | 47 a | 1000 a | 95 d | 8.1 a | 139 a | 133 c | 117 a | 139 a |

Table 4.26.   Comparison of chemical concentration (ppm)
in Red Mangrove (*Rhizophora* sp.) leaves.

| Element | Authority | | |
|---------|-----------|---|---|
| | Rodin and Bazilevich (1967) | Lamberti (1969) | This study Overstory leaves |
| P | 2200 | 1500 | 900 |
| K | 6300 | 9400 | 8400 |
| Ca | 21700 | 14300 | 12200 |
| Mg | 6600 | 5100 | 4700 |
| Na | 22200 | 15700 | 9800 |
| Cu | —— | 4.5 | 6 |
| Fe | *t* | 108 | 82 |
| Mn | 100 | 254 | 387 |
| Zn | | 12 | 11 |

*t*   represents trace amount.

Table 4.27.   Comparison of elemental concentrations (ppm)
in Tropical second-growth.

| Elements | This study | Tergas (1965) | Snedaker and Gamble (1969) | Stark (1970) |
|---|---|---|---|---|
| Phosphorus | 1800 | 1000 | 1600 | 1100 |
| Potassium | 16600 | 11500 | 15900 | 3700 |
| Calcium | 13800 | 6300 | 6000 | 7200 |
| Magnesium | 2500 | 6900 | 7600 | 2500 |
| Aluminum | 470 | —— | 24 | —— |
| Copper | 13 | —— | 17 | 7 |
| Iron | 193 | —— | 132 | 71 |
| Manganese | 174 | —— | 38 | 9 |
| Strontium | 84 | —— | 78 | —— |
| Zinc | 31 | —— | 31 | —— |

and manganese are higher in Panama, while the level of magnesium is lower. The second growth on Surinam podzols has lower concentrations of potassium, copper, iron, and manganese. Calcium levels in Surinam are greater than in Guatemala or Colombia samples, but lower than those in Panama. Phosphorus concentrations are the same as those in Guatemala, while magnesium concentrations are similar to those in Panama. Again, these differences in vegetation probably reflect differences in the soils of the areas.

Since the four second-growth sites were located in an area which was formerly Tropical Moist forest, it also is possible to compare the concentration of elements in early second-growth vegetation and in mature forest (table 4.28). In second growth the abundance relationship be-

Table 4.28.   Average concentration (ppm) of chemical elements in

| Forest | P | K | Ca | Mg | Na | Al | B | Ba | Co |
|---|---|---|---|---|---|---|---|---|---|
| Second-growth | 1800 | 16600 | 13800 | 2500 | 100 | 470 | 18 | 113 | 45 |
| Mature forest | 1200 | 11000 | 17800 | 2000 | 200 | 1050 | 22 | 66 | 41 |

tween elements was potassium > calcium > magnesium > phosphorus > aluminum, while in mature forest the relationship was calcium > potassium > magnesium > phosphorus > aluminum. Several elements such as potassium become less concentrated, while others such as calcium and aluminum, are concentrated with age.

The analyses of the chemical character of forest vegetation have shown that concentration differs by forest type and by compartment for most of the elements examined. Do Panamanian forests also differ in chemical content from other forms of vegetation?

First, the Panama forest concentrations will be compared with those described by Bowen (1966) as typical of angiosperm tissues. The concentrations of the macroelements, phosphorus, potassium, calcium, magnesium, and sodium, are quite similar in both listings (table 4.29), however, tropical forest tissue contains slightly lower concentrations for each of the major elements. In contrast, many of the micronutrients in the tropical forests differ greatly from Bowen's angiosperms. For example, barium, cobalt, cesium, molybdenum, nickel, lead, and titanium are 4 to 100 times greater in tropical forest vegetation than in angiosperm tissue. And concentrations of manganese and zinc are about five times lower in tropical forest vegetation. Not all of these microelements are essential for plant growth and the concentrations in plant tissue may reflect local soil variations. Further, the natural concentrations for some of the elements are low and are difficult to determine accurately. Finally, in some cases Bowen's data are based on samples restricted in area and type of vegetation.

There are relatively few studies of concentrations of chemical materials in mature tropical forest to compare with the data from Panama. Rodin and Bazilevich (1967) provide a list of concentrations in 34 tropical species, mainly from Asia. They also have calculated an average concentration for leaves, trunk and branch wood, and

second-growth forest and mature Tropical Moist forest vegetation.

| Cs | Cu | Fe | Mn | Mo | Ni | Pb | Sr | Ti | Zn | Total ppm |
|----|----|----|----|----|----|----|----|----|----|-----------|
| 27 | 13 | 193 | 174 | 2.2 | 13 | 46 | 84 | 42 | 31 | 36071 |
| 18 | 8 | 176 | 63 | 4.2 | 130 | 36 | 90 | 9 | 27 | 33940 |

Table 4.29.   Comparison of the chemical concentration (ppm) in angiosperms, according to Bowen (1966), and in tropical forest vegetation.

| Element | Angiosperm tissue | Overall mean concentration Panama forest tissue |
|---------|-------------------|-------------------------------------------------|
| P  | 2300 | 1400 |
| K  | 14000 | 12900 |
| Ca | 18000 | 14200 |
| Mg | 3200 | 2500 |
| Na | 1200 | 1000 |
| Al | 550 | 822 |
| B  | 50 | 22 |
| Ba | 14 | 87 |
| Co | 0.48 | 49 |
| Cs | 0.9 | 25 |
| Cu | 14 | 12 |
| Fe | 140 | 267 |
| Mn | 630 | 159 |
| Mo | 0.2 | 3.6 |
| Ni | 2.7 | 49 |
| Pb | 2.7 | 39 |
| Sr | 26 | 97 |
| Ti | 1 | 29 |
| Zn | 160 | 33 |

roots. Ovington and Olson (1970) report on concentrations of Montane forest vegetation in Puerto Rico and Stark (1971a and b) describes elemental concentration in Amazonian forests. In the absence of more specifically applicable data, these summaries provide a basis for evaluations of the tropical pattern of chemical concentration.

The total quantity of elements in the Panama forest biomass averages 3.3 percent dry weight excluding nitrogen, or 4.8 percent including nitrogen. These totals were less than those of Rodin and Bazilevich, who reported concentrations of 4.1 percent, excluding nitrogen, and 5.6 percent, including nitrogen in tropical vegetation. However, their totals included silicon, which made up about one percent dry weight of the total concentration. If silicon occurs at this same concentration in the Panama vegetation, then the total element concentration is approximately the same in both sets of data.

An element by element comparison of the Panama vegetation and the summarized data from Rodin and Bazilevich (1967), Ovington and Olson (1970) and Stark (1971a and b) (table 4.30) showed considerable variation in concentration throughout the forests. For example, in the Panama forest potassium and calcium occur at greater concentrations. In contrast, the Amazonian forests often have the lowest concentration—a point that has been discussed by Stark (1971a and b) and which will be considered later in the concluding sections of this study. Actually for several elements the difference between forests is as great as the differences between sites in Darien. Data on

Table 4.30.  Comparison of chemical concentrations (percent dry weight) in tropical forests. The data represent Panama tropical forest vegetation, overall tropical forests from Rodin and Bazilevich (1967), Montane forest of Puerto Rico (Ovington and Olson, 1970) and selected Amazon forests (Stark, 1971a and b).

| | N | P | K | Ca | Mg | Na | Al | Fe | Mn |
|---|---|---|---|---|---|---|---|---|---|
| Leaves | | | | | | | | | |
| Panama | 2.00 | 0.16 | 1.35 | 1.66 | 0.31 | 0.16 | 0.05 | 0.01 | 0.02 |
| Overall | 2.25 | 0.17 | 0.95 | 1.00 | 0.30 | 0.05 | 0.16 | 0.04 | 0.14 |
| Puerto Rico | 1.60 | 0.78 | 1.04 | 1.01 | 0.37 | 0.20 | —— | —— | —— |
| Amazon | 2.29 | 0.18 | 0.75 | 0.30 | 0.26 | 0.03 | —— | 0.007 | 0.0008 |
| Wood | | | | | | | | | |
| Panama | 0.50 | 0.12 | 0.90 | 0.96 | 0.11 | 0.09 | 0.03 | *t* | 0.01 |
| Overall | 0.34 | 0.04 | 0.25 | 0.47 | 0.10 | 0.01 | 0.04 | 0.03 | 0.02 |
| Puerto Rico | 0.48 | 0.32 | 0.45 | 0.55 | 0.12 | —— | —— | —— | —— |
| Amazon | 0.34 | 0.02 | 0.17 | 0.07 | 0.03 | 0.02 | —— | 0.001 | 0.0002 |
| Roots | | | | | | | | | |
| Panama | 0.80 | 0.11 | 0.66 | 1.36 | 0.26 | 0.17 | 0.13 | 0.05 | 0.01 |
| Overall | 0.65 | 0.08 | 0.45 | 0.60 | 0.17 | 0.02 | 0.15 | 0.15 | 0.03 |
| Puerto Rico | 0.49 | 0.24 | 0.33 | 0.43 | 0.13 | 0.11 | —— | —— | —— |
| Amazon | 1.61 | 0.10 | 0.22 | 0.10 | 0.15 | 0.05 | —— | 0.03 | 0.0009 |
| Total vegetation | | | | | | | | | |
| Panama | 1.40 | 0.14 | 1.29 | 1.42 | 0.25 | 0.10 | 0.08 | 0.03 | 0.02 |
| Overall | 1.22 | 0.10 | 0.82 | 1.14 | 0.28 | 0.38 | 0.15 | 0.12 | 0.11 |
| Puerto Rico | 0.70 | 0.37 | 0.51 | 0.58 | 0.17 | 0.12 | —— | —— | - |
| Amazon | 1.40 | 0.13 | 0.31 | 0.16 | 0.16 | 0.04 | —— | 0.01 | 0.001 |

*t* indicates trace amount.

tropical forest concentrations are inadequate to determine if these site or location differences override tendencies for tropical forests to concentrate specific elements, as Rodin and Bazilevich (1967) suggest.

*Comparisons of soil concentrations.*    The chemical concentrations in forests and soils are exceedingly variable (table 4.31). Within the Tropical Moist forest, iron and zinc concentrations from our study are higher and strontium concentrations are lower than those of other studies. In the Riverine forest, sodium concentrations are low in our study. In Premontane Wet forest, all concentrations from Panama are considerably above those for Costa Rica, except for phosphorus which was lower in Panama. Mangrove mud concentrations are extremely different; in Panama, phosphorus, calcium and manganese were greater than concentrations in Costa Rica, while concentrations of potassium, magnesium and sodium were lower. The data from other types of tropical forests seem equally variable, as does the comparison with world-wide soil concentrations described by Bowen (1966). Indeed, the consistent obvious difference emerging from this comparison is that Premontane Wet forest soils generally contained lower concentrations of macroelements, which presumably reflects the higher rainfall and leaching characteristic of the mountain environment.

We have shown that concentrations of elements in soils and vegetation vary between locations. This variability may be a function of the underlying parent material, as well as topography, rainfall, vegetation, and other environmental factors which interact with the parent material to create microdifferences in the derived soils and differences in species composition and biological processes in the plants. The cause of the variability can be tested by examining the correlation between the concentration of an element in the vegetation and in the soil. The correlation coefficient between the mean concentration in the vegetation and soil of each forest in Panama was greater than -0.70 for calcium, 0.73 for magnesium, and 0.93 for manganese. The coefficient was less than 0.6 for all other elements, suggesting that the concentration in the vegetation is not highly correlated with the total concentration in the soil, except for the three elements mentioned above. The vegetation may reflect available concentrations; however, we need data on exchangeable concentrations of the elements in these soils before we can test this suggestion.

A further way to examine the relationship between chemical concentration in the vegetation and soil is to compare the distribution of abundances of the elements. For example, in mature Tropical Moist

Table 4.31. Elemental concentrations (ppm) in surface soil of tropical forests.

| Forest | Location | P | K | Ca | Mg | Na | Cu | Fe | Mn | Sr | Zn | Exchangeable or total concentration | Authority |
|---|---|---|---|---|---|---|---|---|---|---|---|---|---|
| Tropical Moist | Panama | 2.6 | 41 | 5854 | 521 | 462 | 1 | 6 | 6 | 20 | 10 | Total | This study |
| Tropical Moist | Panama | 7.6 | 117 | 4232 | 645 | 50 | 2 | 6 | 8 | 16 | 51 | Total | This study |
| Tropical Moist | Panama | 9.1 | 650 | 2420 | 700 | — | 1 | 2 | 15 | 214 | 2 | Exchangeable | Gamble et al. (1969) |
| Tropical Moist | Panama | 14.0 | 850 | 4700 | 1620 | — | 1 | 1 | 22 | 72 | 3 | Exchangeable | Gamble et al. (1969) |
| Tropical Moist | Panama | 1.7 | 606 | 7880 | 1144 | — | — | — | — | 55 | — | Exchangeable | Blue et al. (1969) |
| Tropical Moist | Panama | 0.7 | 80 | 7500 | 1330 | — | 1 | 2 | 1 | 251 | 3 | Exchangeable | Gamble et al. (1969) |
| Tropical Moist | Costa Rica | 1.1 | 82 | 1534 | 600 | 85 | — | — | 19 | — | — | Exchangeable | Holdridge 1971 |
| Riverine | Panama | 3.6 | 76 | 1972 | 574 | 673 | 3 | 5 | 36 | 10 | 2 | Total | This study |
| Riverine | Panama | 6.3 | 103 | 4400 | 600 | — | 1 | 1 | 5 | 235 | 3 | Exchangeable | Gamble et al. (1969) |
| Riverine | Panama | 1.6 | 389 | 6710 | 1513 | — | — | — | — | 81 | — | Exchangeable | Blue et al. (1969) |
| Riverine | Costa Rica | 2.4 | 98 | 1120 | 379 | 58 | — | — | 146 | — | — | Exchangeable | Holdridge (1969) |
| Premontane Wet | Panama | 0.2 | 69 | 466 | 302 | 104 | 5 | 58 | 176 | 2 | 2 | Total | This study |
| Premontane Wet | Costa Rica | 0.9 | 27 | 246 | 71 | 51 | — | — | 5 | — | — | Exchangeable | Holdridge (1971) |
| Mangrove | Panama | 24.8 | 17 | 1947 | 42 | 332 | 4 | 102 | 61 | 28 | 13 | Total | This study |
| Mangrove | Costa Rica | 9.0 | 1174 | 1020 | 1560 | 2944 | — | — | 5 | — | — | Exchangeable | Holdridge (1971) |
| Rain forest | Surinam | 113 | 500 | 900 | 1400 | 40 | 9 | 13500 | 3 | — | — | Total | Stark (1970) |
| Second-growth | Guatemala | 5.2 | 195 | 3440 | 1824 | — | — | — | — | — | — | Exchangeable | Tergas (1965) |
| Deciduous Diptccarp | Thailand | — | 140 | 576 | 383 | 39 | — | — | — | — | — | Exchangeable | Tsutumi et al. (1967) |
| Dry Evergreen | Thailand | — | 86 | 120 | 22 | 25 | — | — | — | — | — | Exchangeable | Tsutumi et al. (1967) |
| Moist Tropical | Ghana | — | 109 | 360 | 133 | — | — | — | — | — | — | Exchangeable | Nye and Greenland (1964) |
| Moist Deciduous | India | — | 328 | 2940 | 672 | — | — | — | — | — | — | Exchangeable | Singh, (1968) |
| Wet Evergreen | India | — | 1271 | 2340 | 780 | — | — | — | — | — | — | Exchangeable | Singh (1968) |
| Semi Evergreen | India | — | 702 | 2620 | 492 | — | — | — | — | — | — | Exchangeable | Singh (1968) |
| Soils | World-wide | 650 | 400 to 30000 | 7000 to 500000 | 600 to 6000 | 750 to 7500 | 2 to 100 | 7000 to 550,000 | 100 to 4000 | 50 to 1000 | 10 to 300 | Not indicated | Bowen (1966) |

forest vegetation, the distribution of concentration is Ca > K > Mg > P > Al > Fe = Na > Sr = Mn = Ba. In the soils, the distribution is Ca > Mg > K = Na = Mn > Zn = Sr = Fe > P. The first four elements follow the ratio of exchangeability for montmorillonite clay, with calcium being least replacable (Rankama and Sahama, 1950). Apparently, the vegetation accumulates potassium, phosphorus, and iron or, alternately, discriminates against magnesium, sodium, manganese, zinc, and strontium or both. In the second-growth forests, the distribution of concentrations is very nearly the same as in the mature forest, except that potassium is more abundant than calcium. Riverine and Premontane Wet forests also had distributions similar to Tropical Moist forest for the more concentrated elements. In all forests molybdenum was least abundant.

The explanation of the observed chemical content of forest vegetation must be left in a rather unsatisfactory state. On the one hand, it is clear that the measured chemical concentrations in vegetation and soil are reasonable when compared with similar data from other tropical areas. The macroelements, especially, occur at concentrations reasonably close to other values. However, on the other hand, statistical analyses have revealed numerous differences between and within the forests. These differences are not entirely related to total concentrations in the soils; they also are a function of the vegetation itself. The age of the vegetation is certainly a factor, since younger forests have higher concentrations than mature forests. Further, we have shown that concentrations in vegetation are influenced by stratification of the forest and plant part; lower portions of the forest have higher concentrations and the leaves, fruits, and flowers have higher concentrations than stems. We do not know if the effect of stratification is due to different age of the plant part, different quantities of lichens, algae, and mosses growing on the plant parts, or to physiological differences. Satisfactory explanations of the observed differences between and within forests must await further, more detailed studies.

### THE INVENTORY OF CHEMICALS IN TROPICAL VEGETATION

The total quantities of elements varied greatly in the nine forest ecosystems studied in Panama (table 4.32). Phosphorus, potassium, calcium, sodium, copper, cobalt, iron, manganese, lead, and zinc varied more than a factor of 10 from highest to lowest quantity in a forest type. No one forest type had the maximum levels for the majority of elements.

Quantities of phosphorus, potassium, magnesium, and cobalt were greatest in the Riverine forest; manganese in the Premontane Wet forest; calcium in Rio Sabana Tropical Moist forest; copper in the two-year-old second-growth forest; zinc in the four-year-old second-growth forest; sodium, cesium, iron, lead, and strontium in the Mangrove forest. Considering all the forests, calcium, magnesium, potassium, and sodium were present in greatest quantites (table 4.32). The range in quantities of elements in these Panama ecosystems was 4,000-30,000 kilograms per hectare for calcium, 100-10,000 for magnesium, potassium, and sodium; 10-1,000 for phosphorus, iron, manganese, and zinc; 1-200 for copper, cobalt, cesium, lead, and strontium.

The total quantities of elements described in table 4.32 included those in the vegetation and also in the soil to a depth of 30 centimeters. As mentioned earlier, tropical forests have been characterized as having the largest percentage of the total nutrient capital in the plant compartments, and it is of interest to determine the distribution of elements

Table 4.32.  Inventory of elements (kg/ha) in the soil and vegetation of forest ecosystems in eastern Panama. Values were obtained by multiplying biomass by concentrations. The values are presented unrounded and do not imply accuracy to one kilogram per hectare.

| | Element | | | | | | | | | | | | |
|---|---|---|---|---|---|---|---|---|---|---|---|---|---|
| Forest type | P | K | Ca | Mg | Na | Co | Cs | Cu | Fe | Mn | Pb | Sr | Zn |
| Second growth | | | | | | | | | | | | | |
| 2 yr. July harvest | 72 | 524 | 28705 | 3204 | 1142 | 14 | 6 | 180 | 89 | 232 | 6 | 49 | 19 |
| 2 yr. Oct. harvest | 83 | 565 | 17317 | 2368 | 173 | 11 | 9 | 18 | 61 | 214 | 5 | 65 | 55 |
| 4 yr. July harvest | 154 | 1256 | 17217 | 1985 | 963 | 13 | 5 | 10 | 63 | 459 | 8 | 42 | 411 |
| 6 yr. July harvest | 163 | 903 | 4299 | 1896 | 823 | 9 | 8 | 19 | 56 | 190 | 9 | 52 | 25 |
| Tropical Moist | | | | | | | | | | | | | |
| Rio Sabana | 96 | 1803 | 29251 | 2704 | 2054 | 17 | * | 5 | 42 | 42 | * | 107 | 47 |
| Rio Lara | 274 | 5106 | 22284 | 3267 | 298 | 21 | 15 | 11 | 46 | 78 | 19 | 92 | 235 |
| Premontane Wet | 27 | 2010 | 3939 | 1699 | 867 | 251 | 12 | 24 | 285 | 862 | 12 | 30 | 20 |
| Riverine | 1224 | 10716 | 22085 | 3892 | 3445 | 119 | 21 | 25 | 134 | 182 | 64 | 120 | 38 |
| Mangrove | 439 | 2224 | 12516 | 1632 | 5935 | 51 | 92 | 28 | 713 | 322 | 66 | 190 | 85 |

* Not determined.

within the soil-vegetation system. In the Tropical Moist forest, phosphorus and potassium were the elements mainly concentrated in vegetation. Considering all forests, the percentage of the total quantities present in the plant portion of the ecosystem varied with the age of the forest (table 4.33). Young successional forests contained the smallest fraction of the total nutrients in the vegetation; and, for most elements, the fraction in the vegetation increased with age. Since the concentration of elements was greatest in the youngest stands, the effect shown in table 4.33 is the result of the increase in the biomass with age.

Table 4.33.    Percent of the total inventory contained in the biotic compartments of tropical forest ecosystems in Panama.

| Forest Type | Element | | | | | | | | | | | | |
|---|---|---|---|---|---|---|---|---|---|---|---|---|---|
| | P | K | Ca | Mg | Na | Co | Cs | Cu | Fe | Mn | Pb | Sr | Zn |
| Second growth | | | | | | | | | | | | | |
|   2 yr. July harvest | 54 | 48 | 1 | 1 | >1 | 6 | 12 | >1 | 5 | 1 | 8 | 3 | 3 |
|   2 yr. Oct. harvest | 70 | 63 | 2 | 2 | 3 | 11 | 10 | 2 | 13 | 2 | 16 | 5 | 1 |
|   4 yr. July harvest | 79 | 49 | 3 | 4 | 1 | 21 | 22 | 5 | 15 | 1 | 34 | 5 | >1 |
|   6 yr. July harvest | 82 | 58 | 17 | 5 | 2 | 24 | 16 | 3 | 15 | 2 | 20 | 8 | 6 |
| Tropical Moist | | | | | | | | | | | | | |
|   Rio Sabana | 89 | 89 | 12 | 16 | 2 | 59 | * | 20 | 38 | 38 | * | 19 | 12 |
|   Rio Lara | 88 | 90 | 20 | 13 | 26 | 67 | 53 | 18 | 43 | 55 | 46 | 24 | 4 |
| Premontane Wet | 96 | 85 | 48 | 22 | 47 | 57 | 66 | 7 | 14 | 10 | 75 | 70 | 59 |
| Riverine | 99 | 97 | 61 | 35 | 14 | 92 | 81 | 48 | 84 | 13 | 53 | 63 | 82 |
| Mangrove | 75 | 97 | 32 | 89 | 75 | 53 | 10 | 37 | 37 | 17 | 58 | 35 | 34 |

* Not determined.

For several elements, the mature Tropical Moist forest had a lower percentage concentration of nutrients in the biotic compartments than did the three other mature vegetation types (table 4.33). This pattern is most characteristic for calcium and magnesium. The major exception to the pattern is manganese, which occurred at highest percentage in the vegetation in Tropical Moist forest. The Mangrove forest contained the highest percentages of magnesium and sodium in the biotic compartments, and shared the highest percentage of potassium with the Riverine forest. The Riverine forest contained highest percentages of phosphorus, calcium, cobalt, cesium, copper, iron, and zinc, and the Premontane Wet forest contained the highest percentage of lead and strontium. In this comparison of mature forest systems, there appears to be a direct relationship between the degree of inundation of the soil and the percentage of the community mineral inventory held in the biotic portion of the system. The Mangrove and Riverine forests both are inundated

Table 4.34. Standing crops of selected elements in the vegetation of tropical and temperate forests. Data are in kg/ha.

| Forest type | P | K | Ca | Mg | Na | Vegetation biomass |
|---|---|---|---|---|---|---|
| Tropical forests | | | | | | |
| Tropical Moist (Rio Sabana) | 85 | 1606 | 3502 | 423 | 31 | 276131 |
| Tropical Moist (Rio Lara) | 241 | 4598 | 4702 | 437 | 78 | 377807 |
| Premontane | 26 | 1709 | 1891 | 374 | 407 | 284074 |
| Riverine | 1212 | 10395 | 13472 | 1362 | 482 | 1188760 |
| Mangrove | 329 | 2157 | 4005 | 1452 | 4451 | 469094 |
| Moist Tropical, Ghana[1] | 137 | 906 | 2670 | 390 | —— | 360300 |
| Yamgambi, Congo[2] | 108 | 600 | —— | —— | —— | —— |
| Lower Montane Forest, Puerto Rico[3] | 43 | 517 | 894 | 340 | —— | 262920 |
| Temperate forests | | | | | | |
| Douglas fir forests, Washington[4] | 93 | 259 | 479 | —— | —— | 228311 |
| Spruce-fir, Tennessee[5] | 49 | 221 | 543 | 74 | —— | 325043 |
| Oak Pine, Long Island[6] | —— | 181 | 303 | 44 | 8 | 101920 |
| Pine forest, England[6] | 58 | 180 | 330 | 63 | 15 | 154000 |
| Beech Gap forest, Tennessee[5] | 46 | 189 | 417 | 61 | —— | 169788 |
| *Nothofogus truncata,* New Zealand[7] | 82 | 465 | 1344 | 147 | 32 | 371200 |

[1]Greenland and Kowal (1960).
[2]Bartholomew, et al. (1953).
[3]Ovington and Olson (1970).
[4]Cole, et al. (1968).
[5]Shanks, Clebsch, and DeSelm (1961).
[6]Woodwell and Whittaker (1968).
[7]Ovington (1962).

periodically and the organic compartments contained the largest percentages for many of the 13 elements (table 4.33).

Finally, let us compare the inventory of chemicals in tropical and temperate forests. Tropical vegetation contains larger standing crops of mineral elements than does temperate vegetation (table 4.34). This is partly due to the fact that tropical forests contain a much larger amount of organic biomass; however, as discussed earlier, they also may have a higher concentration per unit of biomass than do temperate forests. A regression comparison of the relationship between the mineral inventory and the vegetation biomass in the tropical and temperate forests shown in table 4.34, following the procedure of Tsutsumi et al. (1968), showed that quantities of phosphorus, potassium, calcium, and magnesium were significantly correlated with biomass while the relationship for sodium was not significant. The correlation coefficients for the elements were: phosphorus 0.9754, potassium 0.9275, calcium 0.9599, magnesium 0.7759, and sodium 0.1676. The regression formulas for the four significant correlations are shown below:

$$
\begin{aligned}
\text{phosphorus} \quad & Y = 212X^{0.0018} \\
\text{potassium} \quad & Y = 1900X^{0.00974} \\
\text{calcium} \quad & Y = 2763X^{0.01248} \\
\text{magnesium} \quad & Y = 436X^{0.00134}
\end{aligned}
$$

where $Y$ is the concentration in the biomass in ppm and $X$ is the biomass standing crop. For these four elements, the amount of vegetation more strongly influences the total inventory, while for an element such as sodium, special forest conditions, such as are found in Mangroves, make a correlation between biomass and chemical inventory insignificant. Tsutsumi et al. (1968), also showed correlations between biomass and nitrogen, potassium and phosphorus in Japanese forests.

# V
# MINERAL CYCLING: CONCLUSIONS AND SUMMARY

The tropical forest has been characterized as an ecosystem containing a large biomass, a large inventory of chemical elements and rapid cycles of minerals between the organic components and the substrate. We have described the mineral cycling in one type of tropical forest, the Tropical Moist forest, which is adapted to alternating wet and dry seasons. In this final section we will summarize the information on the kinetics of the Tropical Moist forest and relate it to mineral cycling of ecosystems generally. But before we discuss the dynamic system, we will briefly consider the static pattern of mineral concentration.

The mineral distribution of forests and forest components has been shown to differ significantly in several contexts. For example, the average concentration of elements in the vegetation was highest in selected second-growth stands of Tropical Moist forest and least in Premontane Wet forest (table 4.18). Second-growth forests contained higher total concentrations than mature Tropical Moist forests (table 4.18). Understory fruits, flowers, and leaves had higher concentrations than other vegetation compartments, while overstory stems had lowest concentrations (table 4.20).

The observed differences between forests were explained by differences in their substrates and environments. Tropical Moist forest is found on a dark clay soil derived from shale in an area of relatively little topographic relief, while Premontane Wet forest occurs on a weathered red soil in a highly dissected mountainous area, which experiences higher rainfall. There also may be differences in the uptake and storage by the individual species which make up the forests but we have no information on patterns of nutrient distribution in species.

Further, it was observed that actively growing tissues such as fruits, flowers, and leaves had higher nutrient concentrations than less active tissues such as wood. Actively growing secondary vegetation also had higher concentrations than mature vegetation. These differences were expected since growing and developing tissues generally have greater nutrient requirements than inactive or mature tissues. It is not clear why understory components of the forests had higher concentrations than overstory components. Presumably the biota living in profusion on the understory leaves and stems are instrumental in concentrating minerals in these strata of the forest.

Considering the elements individually, calcium and potassium were most abundant in the vegetation of all forests combined, and magnesium, phosphorus, sodium, and aluminum were next most abundant. The ratios of abundance in the vegetation did not follow those of the soils where calcium occurred at highest concentrations, followed by magnesium, sodium, potassium, and phosphorus. These relationships are partly controlled by the geology of the region and partly by the requirements of the biota. For example, the Sabana shales are especially rich in calcium and magnesium, and the reflection of the underlying rock in the soils and vegetation is one reason for the differences in the mineral concentrations in Darien forests and those of the Amazon (Stark, 1971a and b) and Puerto Rico (Ovington and Olson, 1970) shown in table 4.30.

A comparison of the rate of cycling of elements from Tropical Moist forest vegetation to the soil with the inventory in the soil allows a grouping of the elements into four categories: 1) those with a small soil inventory and rapid rate of cycling, 2) those with small inventory and slow cycling rate, 3) a large soil inventory and rapid rate, and 4) a large soil inventory and slow cycling rate. Elements in group one would be expected to be limiting since the demand of the biota may not be able to be fulfilled from the soil reservoir. Those in group three might be limiting depending upon the magnitudes of the rates and inventory. Elements in groups two and four are less likely to be limiting to biological production.

Phosphorus is the only element falling in group one and potassium is the only element in group three (table 5.1). Calcium, magnesium, and sodium are in group four; all remaining elements are in group two. Thus, we conclude that of the elements studied phosphorus and potassium are the two that may be most critical. However, since input and output of phosphorus and potassium from the forest ecosystem is balanced (table 3.20), these elements probably are not limiting under the present conditions.

The above comparison considers the vegetation operating dynamically while the soil inventory is held constant. Since the soil system also is dynamic, another useful comparison is to contrast the movement between the vegetation and soil with the flux from the soil to the streams. This latter flow is a measure of the rate of geochemical cycling. Phosphorus and potassium cycle between the vegetation and soil at rates 17 to 21 times greater than from the soil to streams (table 5.1). For calcium, cobalt, copper, manganese, strontium, and zinc the biological cycle is equal to or only one to three times the geochemical

Table 5.1.   Summary of mineral cycling dynamics
in Tropical Moist forest.

| Element | Soil inventory kg/ha | Annual loss from vegetation to soil kg/ha/yr | Loss as percent of inventory | Discharge from soil kg/ha/yr |
|---------|--------|--------|--------|--------|
| P  | 22    | 12   | 54.5  | 0.7   |
| K  | 353   | 197  | 55.8  | 9.3   |
| Ca | 22166 | 298  | 1.3   | 163.2 |
| Mg | 2256  | 35   | 1.6   | 43.6  |
| Na | 1121  | 27   | 2.4   | 92.5  |
| Co | 7     | 1.2  | 17.1  | 0.7   |
| Cu | 7     | 0.4  | .1    | 0.4   |
| Fe | 26    | 4.9  | 18.8  | 10.1  |
| Mn | 31    | 0.8  | <.1   | 0.3   |
| Sr | 79    | 1.1  | 1.4   | 0.4   |
| Zn | 134   | 0.9  | <.1   | 0.6   |

flux. For magnesium, sodium, and iron the geochemical cycle is greater than the biological cycle. Again, of the elements studied, phosphorus and potassium are the most critical to the system.

Finally, we can compare the movement of chemicals from the vegetation to the soil with the chemical inventory in the vegetation. This is a relatively common comparison in the literature since data are often collected on litter fall in forests and on the standing crops of elements in the vegetation (for example, see Rodin and Bazilevich, 1967) yet it has a fundamentally different meaning than the above comparisons because it considers the vegetation as the reservoir of minerals rather than the soil. Sodium and the microelements cobalt, copper, iron, and zinc were lost in greatest percentage amounts from the vegetation (table 3.11). Over 50% of the sodium and iron in the above ground biomass turns over annually. These data illustrate the rapidity of the turnover of microelements in the biotic compartments and suggest that these minerals turn over at a relatively faster rate than those which are most limiting in the soil.

Input to the Tropical Moist forest from rainfall almost balances output in stream discharge for most elements examined in the study (table 3.20). Large differences between input and output were observed only for calcium, magnesium, and sodium. If the forest cycles are in equilibrium as we have assumed, then these deficits must be made up from the soil system, presumably through the weathering of soil minerals. Turnover time for the total soil and vegetation inventory is more than 100 to 200 years for phosphorus, potassium, calcium, manganese, strontium, and zinc (table 3.20) and less than 100 years for magnesium, sodium, cobalt, copper, iron, and lead.

Table 5.2.   The relation of rain input of elements
to the biotic inventory of mature Tropical Moist forest (kg/ha).

| Element | Biotic inventory | Rain input | Years of input to equal inventory |
|---|---|---|---|
| P | 164 | 1 | 164 |
| K | 3103 | 9.5 | 327 |
| Ca | 4103 | 29 | 141 |
| Mg | 429 | 4.9 | 88 |
| Na | 55 | 31 | 2 |
| Co | 12 | 2 | 6 |
| Cu | 2 | 0.5 | 3 |
| Fe | 18 | 3 | 6 |
| Mn | 29 | 0.4 | 73 |
| Sr | 21 | 0.1 | 210 |
| Zn | 8 | 0.9 | 9 |

If the forest was removed the organic inventory which we have been discussing would be wholly or partially lost and would need to be replaced when new vegetation was established. Disregarding change in the nutrient output from the system which would be expected to increase with the destruction of the forest, it would require several hundred years to rebuild the organic nutrient inventory solely from the rain input (table 5.2). Since succession on small plots proceeds rapidly, the soil reservoir is obviously adequate for replacement of the required nutrients. However, if the area of disturbance was very large the capacity of the soil to continue to supply the quantities of chemicals required may be limited. The above comparisons show that potassium and phosphorus are the two elements most likely to be limiting to forest regrowth. Both occur in relatively large amounts in the organic inventory (table 5.1), both cycle rapidly through the system, and phosphorus occurs in relatively small amounts in the soil. Further, input and output from the forest ecosystem is balanced for these elements.

These conclusions should not be interpreted to mean that intensive agriculture or forestry could not be carried out in Darien Province. Rather, they suggest that replacement of the natural forest with other vegetation should be carried out in such a way as to avoid nutrient loss from the soil reservoir. Then, in addition, detailed studies of changes in soils under different length of cropping and different types of management are needed. Krebs (1972) has carried out this type of study in Costa Rica and concluded that deterioration of soil fertility can be reversed by proper management and fertilization. Actually, as mentioned in the Introduction, there is evidence that the Darien supported relatively large populations of Indians who maintained extensive fields in the pre-Columbian period. Several ecologists, including Budowski (1966), who are acquainted with the history of the area have concluded that the Tropical Moist forest is successional to these Indian fields and to open lands maintained by burning. It would be exceedingly interesting to know how large the areas of cultivation actually were and the methods of agriculture used by the Indians. An agro-ecological experiment station based upon the indigenous methods of cultivation could be organized in Darien Province with considerable profit to the inhabitants.

The operation and maintenance of the forest mineral cycling mechanisms described in this report require a continuous source of energy. In the forest energy for plant growth and development is obtained through organic primary production. Generally tropical forests have high rates of production (Golley, 1972a and b). In the Tropical Moist forest we measured only the production of litter (table 3.1), which amounted to 11,350 kg/ha/yr. If we add to the litter or leaves the production of roots, stems, and fruits, the total net production probably would fall in the range for tropical forests of 15 to 20 metric tons/ha/yr, reported by Golley (1972a). Assuming that net production is 30 percent of gross primary production, and that most of the photosynthate is used in forest maintenance rather than growth (Golley, 1971b), then gross primary production might be as high as 50 metric tons/ha/yr. Since the caloric value of tropical forest vegetation is about 3,900 gram calories per gram dry weight (Golley, 1969), this level of gross primary production would represent a power input of about $195 \times 10^3$ kcal/ha/yr. Considering that a temperate grass field in the United States may have a power input of $30 \times 10^3$ kcal/ha/yr (Golley, 1965), this is a substantial power source for dynamic forest processes.

The energy stored in the forest biomass (table 5.3) equals almost $17 \times 10^8$ kcal/ha/yr or about 10 times the annual power input. This stored energy and that derived from photosynthesis is not the sole energy source driving the mineral dynamics. As Odum (1970) has pointed out, the actual movement of water carrying the essential nutrients into and through the system operates on net radiation energy and not on photosynthetic energy. These two power sources, net radiation employed in evaporation and transpiration and energy converted through photosynthesis, together provide the energy for cycling of minerals and other ecosystem functions.

Table 5.3.  Energy storage in the Tropical Moist forest.
Data on energy values from Golley, 1969.

| Component | Energy value Kcal/gram | Energy storage Kcal/ha $\times 10^6$ |
|---|---|---|
| Overstory leaves | 3.9 | 44 |
| Overstory stems | 4.2 | 1489 |
| Understory leaves | 3.6 | 2 |
| Understory stems | 4.2 | 5 |
| Fruits and flowers | 4.3 | 1 |
| Roots | 4.0 | 39 |
| Total living | —— | 1580 |
| Litter | 4.1 | 72 |
| Total living and dead | —— | 1652 |

PATTERNS OF MINERAL CYCLING

In textbooks mineral cycling is often discussed as a single dynamic process analogous to energy flow. In actuality the concentration and inventory of elements in the forest ecosystem is a complex process with each element varying as a function of its physical chemistry, geochemistry, and biochemistry. The physical characteristics of the element concerned with its nuclear stability, reactivity, and size are fundamental to an explanation of the forest chemistry. However, the elements are distributed in the ecosystem in a variety of chemical combinations and not as elements alone, and these compounds are the result of geochemical actions such as rock formation, rock weathering, soil formation, and soil leaching. Further, the plants and animals concentrate, accumulate, and discriminate against the elements in the lithosphere, atmosphere, and hydrosphere. The distribution and cycling of specific elements can be explained through the interaction of these three sets of chemical phenomena. For example, phosphorus has been identified as a limiting element to the forest system. Phosphorus is chemically relatively rarer than its neighboring elements in the periodic table since it has an odd atomic number (Oddo and Harkin's Law, see Rankama and Sahama, 1950), it has a relatively low abundance in igneous rock and in shale, but it is an essential element in living organisms. On the other hand, calcium is an even atomic number element, is abundant in the lithosphere, and is an essential component of the chemistry of cell walls. In montmorillonite clays calcium is preferentially fixed in the presence of sodium (Rankama and Sahama, 1950). Thus, it is not surprising that phosphorus may be critical in the Tropical Moist forest and that calcium is the most abundant element in the forest biomass.

In investigating the chemistry of the tropical forests we had two research strategies available to us in 1966. First, our studies could have concentrated on the movement of one or a few elements through the ecosystem focusing on the flux between forest components or, second, we could investigate as large a suite of elements as our analytical capabilities would permit with less emphasis on flux data. Since Battelle Memorial Institute required information on the elements actually present in the forest we adopted the second approach. At that time we concluded that it was unlikely that we could extrapolate from detailed information on a single element to the distribution or cycling of other elements in the system. Our strategy seemed to be sound since we have been unable to determine a constant pattern of correlation between the elements in such system compartments as the vegetation and soil in any of the forests and we have found a number of statistical differences in element content of forests and components. However, recently it has been suggested that the statistical technique of factor analysis (Comrey, 1973) may be used to determine interrelationships between groups of elements. We anticipate that with larger sets of data on a greater variety of chemicals, factor analysis might permit the separation of physical, geochemical, and biochemical effects, and allow a more rigorous explanation of the observed patterns of mineral distribution and movement.

Given the complexity of the process and fairly extensive data on the standing crops of chemical elements in the forest ecosystem, we can explore several other patterns of mineral cycling in these forests. First, there is a well known ecological hypothesis or generalization which states that tropical forests have evolved mechanisms to conserve essential minerals as a response to high rates of decomposition and water flux through the system. This generalization can be tested with two sets of data from this study. First, the percentage of the total mineral pool held in the vegetation indicates the amount of the inventory under biological control and conserved against the erosive forces of the environment. Second, the ratio of the rate of biological cycling to the rate of geological cycling indicates the extent of biological control over the mineral flux of the system. A large biological flux and a low geological flux indicate tight cycling with a relatively low rate of leakage from the forest.

We have shown that more than 80 percent of the phosphorus and potassium was located in the biomass in each of the mature forests studied in Darien Province (table 4.34). The pattern of distribution was much more ambiguous for the other elements. For example, over 50 percent of the inventory of calcium, cobalt, cesium, lead, strontium, and zinc occurred in the biomass in several forests, but not in all forests. Further, Riverine and Mangrove forests appeared to contain

Table 5.4.   Ratio of the rate of biological uptake of elements to output in stream water in Tropical Moist forest.

| Element | Ratio |
| --- | --- |
| P | 15.0 |
| K | 20.2 |
| Ca | 1.7 |
| Mg | 0.7 |
| Na | 0.03 |
| Co | 0.5 |
| Cu | 0.3 |
| Fe | 0.2 |
| Mn | 1.2 |
| Pb | 0.2 |
| Sr | 2.5 |
| Zn | 0.1 |

a higher percentage of elements in the biomass than did Tropical Moist or Premontane Wet forests. Also, as mentioned earlier, the biological cycle of phosphorus and potassium was much greater than the geological flux in the Tropical Moist forest (table 5.4), but for the other elements the cycles were more nearly similar or the geological flux was greater.

We conclude that phosphorus and potassium, as critical elements in Tropical Moist forest, are conserved by incorporation in the biotic part of the system and by rapid reincorporation in the biomass when released from litter and soil. Supplies and losses of other elements are less critical and mineral conservation adaptations are not obvious. Of course, this does not mean that the individual species will not conserve specific elements. The competition between species for essential nutrients may be of considerable importance in the distribution and survival of populations in these forests.

We suspect that this conclusion regarding the mineral conservation hypothesis is specific to the Tropical Moist forest. We would expect to find substantial differences in the patterns of mineral cycling in different types of forests, since within one relatively small area, Darien Province, the forests differed significantly in their chemical content and in the proportion of the mineral inventory held in the living part of the system. Mineral cycling probably varies with the nutrient supply to the system, with the time available for the system development on the site, and with environment. For example, the soils in certain Amazonian forests (Stark, 1971a and b; Klinge and Rodrigues, 1968a and b) are podsols which contain low quantities of nutrients and have low exchange capacity (Stark, 1971a). As a consequence, concentrations of nutrients in plants are much greater than in the soil (Stark, 1971a) and elements released from litter are rapidly taken up by vegetation and do not appear in forest stream waters (Brinkman and dos Santos, 1971). The litter also is low in nutrients compared to that of other forests (Klinge and Rodrigues, 1968b). The Amazon case has led Stark (1971a) to propose a direct nutrient cycling hypothesis which states that on poor Amazon sands nutrients are transported from dead organic matter by mycorrhizal fungi directly to living plant roots without appearing in the soil solution. Given adequate time and freedom from disturbance lush tropical forests have developed on these Amazon lake sediments, but Stark suggests that in time the soil of the root zone will become entirely depleted of nutrients and slow leakage from the biological mineral cycle will reduce the nutrient reserve in the vegetation, resulting in loss of rooted plants and renewal of the erosion cycle. Obviously cutting the forest will only hasten the process.

Misra (1972) presents data on another extreme case within the continuum of tropical moisture conditions, that of the tropical dry deciduous forest. The annual rainfall where this forest grows in India is 700 to 800mm and is restricted to a short season. The total biomass of the forest is 239 metric tons/ha and 44 percent of the phosphorus, 23 percent of calcium, and 15 percent of the nitrogen is held in the living organic part of the system. However, turnover of the nutrients in the biomass is rapid (P, 44 years; Ca, 20 years; and N, 70 years). Apparently in this environment availability of water is the major limiting factor to forest production. Because of the environmental conditions the amount of nutrients that can be incorporated into the biomass is relatively small and these are conserved by rapid cycling between the biomass and substrate.

These several examples of mineral standing crops and cycling in tropical forests suggest a restatement of the mineral conservation hypothesis described earlier. On the site with optimum temperature and moisture conditions, with availability of the appropriate biota and adequate time a complex multistrata tropical forest can develop. Since the environmental conditions for optimum production are also those for rapid chemical reaction and leaching from the soil, it is not surprising that high production, large proportionate storage in the biomass, rapid mineral cycling and high rates of weathering and leaching of the soil are all related. Of course, storage of essential nutrients in the biomass and high rates of litter fall and decomposition are an advantage to the community but these seem to be consequences of environment and primary production rather than specific features of the vegetation selected through evolution.

Sites with less than optimum conditions for growth might have poor nutrient status in the soil (the podzolic sands in Amazonia), low temperature (in mountainous areas), low water availability (the dry deciduous forest in India), or a combination of these factors. Where the nutrient status is suboptimal multistrata forest can develop if appropriate species and sufficient time for development are available. In Darien rain input of nutrients alone would be sufficient to rebuild the nutrient inventory over several hundred years (table 5.2) assuming that the forest has mechanisms to retain these nutrients in the system. Such mechanisms might include the mycorrhizal-root linkages found by Stark (1971a) or the movement of nutrients from dying leaves or branches to the more active tissues before they fall to the ground as litter.

In a situation where temperature or water is limiting to growth, the tropical forest responds in a way similar to forests in a temperate environment. Growth is limited to a certain period and the biomass and mineral cycling curtailed accordingly. The available examples offer a tantalizing glimpse of the array of possible ecosystem adaptations in the tropical environment. Obviously, we require a much larger set of observations before we can describe the full range of adaptations.

The second pattern to consider concerns the stability of the forest chemical kinetics. While the study was not designed to determine system stability, it might be useful to speculate about the control processes which could maintain the assumed ecosystem steady state. For example, control might be achieved through a large mass which is relatively insensitive to short term perturbations. The simulation study showed that the stems or woody tissues functioned to damp perturbations in the system (fig. 3.9) and, thus, the stems not only are involved in mineral conservation through the storage of nutrients but also may act in the control of cycling. Alternatively, a small mass operating on a large flux can alter the flux by relatively small changes in the size of the component. We have seen how the system is sensitive to the mass of detritivores; other animals also probably play an important role in controlling the rate of specific cycling pathways in this way (Golley, 1971). This is an area of fruitful research and further studies could be profitably focused on this problem.

Besides these features of system maintenance, if the ecosystem is destroyed succession or recovery takes place with eventual reestablishment of a steady state. A number of conditions must be present for recovery to occur, among these are the presence of adequate supplies of available nutrients. Second growth species may have special adaptations such as deep feeding roots, special accumulation mechanisms, or unusual storage capacities to assure that nutrients can be collected from the soil profile and stored. In the Tropical Moist forest succession on shifting agricultural plots was quite rapid, with reestablishment of the leaf area by the sixth year after abandonment (fig. 5.1). As mentioned earlier, the rapidity of recovery of the vegetation in this forest suggests that nutrients probably are not limiting. However, we suspect that the rate of recovery of tropical forest vegetation depends on the extent and nature of the disturbance just as it does in temperate regions (Golley, 1965). Small shifting agricultural plots can be resupplied with nutrients from surrounding forests and soils. Plant roots move rapidly into undisturbed boundary areas and a nutrient deficiency is avoided. In contrast, large areas of deforested

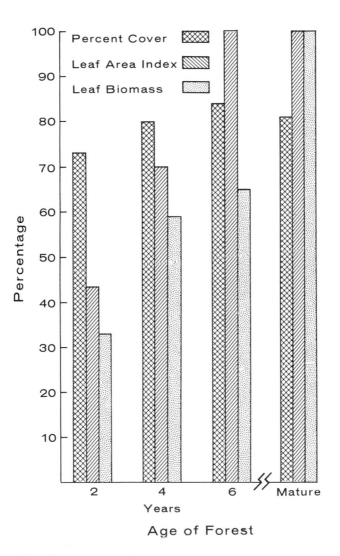

Figure 5.1 Comparison of vegetation characteristics in three second growth forests and in mature forest. The percent cover is compared to percent leaf area index and leaf biomass of the mature forest.

land will probably exhibit much slower rates of revegetation. Indeed, where all mature forest is destroyed and natural vegetation is entirely secondary the former primary forest may never be reestablished.

This study was initiated in 1966 before many ecosystem investigations were in operation. Since that time the International Biological Program (IBP) has supported a variety of studies which will provide data on mineral cycling in ecosystems, although very few of these studies are in the tropics. The success of the IBP both from a theoretical, as well as a practical view has led to plans for further expansion of ecosystem studies after IBP terminates in 1974-75. Unesco Man and the Biosphere, Instituto Interamericano de Ciencias Agricolas tropics program, and the Institute of Ecology (TIE) tropical ecology workshop have encouraged productivity and mineral cycling investigations especially in tropical forests and grasslands. Considering our experience in Tropical Moist forest, what general guidelines can be drawn for these future studies?

First, the area of investigation should encompass a watershed or similar landscape unit. Smaller areas may not be representative of the landscapes, contain sufficient *in situ* variation, or allow the measurement of the flux of water and chemicals from the system. One of the major objectives of a study should be the comparison of the biological and geochemical fluxes and the role of the biotic community in regulating system output. To do this it is necessary to determine the movement of water and chemicals through the soil strata or through stream discharge on several watersheds or on different portions of a landscape unit. In this way differences in system behavior can be identified, or alternately, the biotic community could be manipulated and system input and output monitored. Next, the system can be dissected and those components important to the particular process (mineral cycling, energy flow, productivity, stability) can be studied individually in the context of an explanation of whole system properties. In this type of study there is no point in making a total species list or attempting to determine the properties of all species with the view of adding these together to obtain the total system behavior. This approach is impractical and may well be impossible, especially in a tropical forest where the number of tree species alone is in the hundreds. This is not to say that studies of species populations are unimportant. But from the point of view of explaining the behavior of the entire system, study of all species is not a reasonable research strategy. These ideas underlay the development of the present study and in retrospect we feel that the approach was sound and allowed us to meet our objectives.

# APPENDIX

# Plant Species in the Forests of Darien, Panama

JAMES A. DUKE

*Battelle Memorial Institute*

Tropical forests with their large number of plant species are poorly known botanically because they have not been as thoroughly collected as the more accessible and less complicated types of vegetation in other parts of the world. The paucity of taxonomic information and the difficulty of making identifications required ecologists to consider the forests from a physiognomic and geographic rather than a taxonomic point of view. Even so, the ultimate aim of ecological studies is to specify the function of each species population in the overall processes of the community ecosystem. Therefore, it was important to provide at least a preliminary taxonomic description of the forest types to serve as a foundation or guide for further work.

For this purpose, I visited the harvest plots where forest structure was being studied and traveled extensively through each forest type recognized by the ecologists. Plant species which were collected, observed, or expected are listed in the following tables. The species are categorized on the basis of forest type, the strata of the forest in which they occur, and their relative abundance. New and unpublished species are in parentheses. The abundance symbols are: *d*—locally dominant, *f*—frequent, *o*—occasional, *r*—rare, *u*—insufficiently known to assign an abundance rating. While only about 50 percent of the determinations have been received, the lists provide another and different, though incomplete, description of the forest communities. The number of identified plant species was greatest in the Tropical Moist (399 taxa) and least in the Mangrove forest (19 taxa). The other forests were similar to Tropical Moist forest: Premontane Wet (382), Second-growth (299), and Riverine forest (218). The number of canopy and sub-canopy trees was greatest in Tropical Moist (208 taxa) and Premontane Wet forest (180 taxa). Riverine forest (66), Second-growth (38), and Mangrove (8) had substantially few identified canopy and sub-canopy trees.

Appendix table 1.    List of species in Riverine forest.

| Compartments | Relative abundance |
|---|:---:|
| Understory herbs | |
| *Eclipta alba* (L.) Haask. | F |
| *Justicia comata* (L.) Lam. | F |
| *Panicum maximum* Jacq. "yierba guinea" | F |
| *Priva lappulacea* (l.) Pers. "cadillo" | F |
| *Scoparia dulcis* L. "escobilla amarga" | F |
| *Tibouchina longifolia* (Vahl) Baill. | F |
| *Axonopus compressus* (Sw.) Nees | O |
| *Borreria laevis* (Lam.) Griseb. | O |
| *Chaptalia nutans* (L.) Polak | O |
| *Chrysothemis friedrichsthaliana* (Hanst.) Moore | O |
| *Crinum darienensis* Woodson | O |
| *Dorstenia contrayerba* L. "contrayerba" | O |
| *Drymaria cordata* (L.) Willd. | O |
| *Echinodorus longipetalus* Micheli | O |
| *Eleocharis nodulosa* (Roth.) Schult. | O |
| *Fleurya aestuans* (L.) Gaud. "ortiga" | O |
| *Hydrocotyle mexicana* Schlecht. and Cham. | O |
| *Hyptis verticillata* L. "paleca" | O |
| *Lindernia crustacea* (L.) F. Muell. | O |
| *Panicum grande* Hitchc. | O |
| *P. zizanioides* H.B.K. | O |
| *Paspalum paniculatum* L. | O |
| *P. plenum* Chase | O |
| *P. plicatulum* Michx. | O |
| *P. virgatum* L. "cabezona" | O |
| *Setaria vulpiseta* (Lam.) R. and S. | O |
| *Spigelia anthelmia* L. | O |
| *Beloperone variegata* Lindau | R |
| *Cyperus esculentus* L. | R |
| *Cyperus ferax* L. C. Rich. | R |
| *Cyperus luzulae* (L.) Retz. | R |
| *Fimbristylis miliacea* (L.) Vahl | R |
| *Sauvagesia erecta* L. | R |
| Understory Bromeliads, Palmitas, and Platanillas | |
| *Bactris major* Jacq. "mongo lolo" | D |
| *Corozo oleifera* (H.B.K.) Bailey "corozo" | D |
| *Gynerium sagittatum* (Aubl.) Beauv. "caña blanca" | D |
| *Renealmia cernua* (Sw.) MacBride | F |
| *Aechmea magdalenae* (Andre) Andre ex Baker "pita" | O |
| *Bromelia karatas* L. "pita" | O |
| *Bromelia pinguin* L. "pinuela" | O |

Appendix table 1.  List of species in Riverine forest, *cont.*

| Compartments | Relative abundance |
|---|:---:|
| *Calathea allouia* (Aubl.) Lindl. "bijao" | O |
| *Calathea insignis* Peters "bijao" | O |
| *Carludovica palmata* R. & P. "portorico" | O |
| *Costus villosissimus* Jacq. "cañagria" | O |
| *Cyclanthus bipartitus* Poit. "portorico" | O |
| *Heliconia mariae* Hook. F. "platanilla" | O |
| *Maranta arundinacea* L. "sagu" | O |
| *Renealmia aromatica* (Aubl.) Griseb. | O |
| *Thalia geniculata* L. | O |
| *Xanthosoma helleborifolium* (Jacq.) Schott | O |
| *Xiphidium caeruleum* Aubl. "palmita" | O |
| *Phytelephas seemannii* Cook "tagua" | R |
| Understory shrubs | |
| *Lindenia rivalis* Benth. | D |
| *Cassia reticulata* Willd. "laureño" | F |
| *Chiococca alba* (L.) Hitchc. "lágrimas de Maria" | F |
| *Clibadium leiocarpum* Steetz "catalina" | F |
| *Gonzalagunia rudis* Standl. "nigüita" | F |
| *Hasseltia floribunda* H.B.K. "parimontón" | F |
| *Henrietella fascicularis* (Sw.) Tr. | F |
| *Lantana velutina* Mart. and Gal. | F |
| *Leandra dichotoma* (Don) Cogn. | F |
| *Mabea occidentalis* Benth. | F |
| *Mimosa pigra* L. "dormilon" | F |
| *Neurolaena lobata* (L.) R.Br. "contragavilán" | F |
| *Palicourea guianensis* Aubl. | F |
| *Palicourea triphylla* DC. | F |
| *Psychotria horizontalis* Sw. | F |
| *Ricinus communis* L. "higuerillo" | F |
| *Siparuna guianensis* Aubl. "pasmo" | F |
| *Triumfetta lappula* L. "abrojo" | F |
| *Acalypha macrostachya* Jacq. | O |
| *Aphelandra micans* Moritz | O |
| *Bixa orellana* L. "achiote" | O |
| *Cassia fruticosa* Mill. | O |
| *Cassia oxyphylla* Kunth | O |
| *Clidemia densiflora* (Standl.) Gleason | O |
| *Clidemia dependens* D. Don | O |
| *Clidemia septuplinervia* Cogn. | O |
| *Conostegia micromeris* Standl. | O |
| *Cordia ferruginea* (Lam.) R. and S. | O |
| *Coutarea hexandra* (Jacq.) Schum. | O |

Appendix table 1. List of species in Riverine forest, *cont.*

| Compartments | Relative abundance |
|---|:---:|
| *Hamelia axillaris* Sw. | O |
| *Malvastrum americanum* (L.) Torr. | O |
| *Malvaviscus arboreus* Cav. "papito de monte" | O |
| *Miconia lacera* (Bonpl.) Naud. | O |
| *Miconia lateriflora* Cogn. | O |
| *Miconia micromeris* Standl. | O |
| *Miconia nervosa* (Sm.) Tr. | O |
| *Mouriri parviflora* Benth. | O |
| *Muntingia calabura* L. "periquito" | O |
| *Myriocarpa yzabalensis* (Donn. Sm.) Killip | O |
| *Ossaea diversifolia* (Bonpl.) Cogn. | O |
| *Piper aduncum* L. "cordoncillo" "gusanillo" | O |
| *Piper grande* Vahl | O |
| *Piper tuberculatum* Jacq. "gusanillo" | O |
| *Piper villaramularum* C. DC. | O |
| *Psychotria rufescens* H. and B. | O |
| *Randia aculeata* L. | O |
| *Randia mitis* L. | O |
| *Rustia occidentalis* (Benth.) Hemsl. | O |
| *Solanum rugosum* Dunal | O |
| *Solanum umbellatum* Mill. | O |
| *Tournefortia obscura* A. DC. | O |
| *Urera elata* (Sw.) Griseb. | O |
| *Biophytum sp.* | R |
| *Codiaeum variegatum* (L.) Blume "croton" | R |
| *Dicraspidia donnel-smithii* Standl. | R |
| *Henrietella fissanthera* Gleason | R |
| *Miconia cionotricha* Uribe | R |
| *Miconia oinochrophylla* Donn. Sm. | R |
| *Montrichardia arborescens* (L.) Schott "castaño" | R |
| Woody vines | |
| *Byttneria aculeata* Jacq. "espino hueco" | F |
| *(Mucuna bracteata* Dwyer) "ojo de venado" | F |
| *Phryganocydia corymbosa* (Vent.) Bur. | F |
| *Strychnos panamensis* Seem. "canjura" | F |
| *Wulffia baccata* (L.f.) Kuntze | F |
| *Adelobotrys adscendens* (Sw.) Triana | O |
| *Adenocalymma* aff. *inundatum* Mart. ex DC. | O |
| *Allamanda cathartica* L. | O |
| *Arrabidaea panamensis* Sprague | O |
| *Combretum fruticosum* (Loefl.) Stuntz | O |
| *Connarus panamensis* Griseb. | O |

Appendix table 1.    List of species in Riverine forest, *cont.*

| Compartments | Relative abundance |
|---|---|
| *Entada gigas* (L.) Fawc. and Rendle | O |
| *Machaerium seemannii* Benth. | O |
| *Merremia umbellata* (L.) Hall f. | O |
| *Omphalea diandra* L. | O |
| *Prestonia portobellensis* (Beurl.) Woods. | O |
| *Vitis tilifolia* H. and B. "uva" | O |
| *Bauhinia standleyi* Rose | R |
| *Cheiloclinum cognatum* (Miers) A. C. Sm. | R |
| *Hiraea faginea* (Sw.) Ndzu. | R |
| Herbaceous vines | |
| *Eupatorium iresineoides* H.B.K. | F |
| *Eupatorium macrophyllum* L. | F |
| *Eupatorium microstemon* Cass. | F |
| *Eupatorium odoratum* L. | F |
| *Gurania coccinea* Cogn. "ya te vi" | F |
| *Mikania guaco* H.B.K. | F |
| *Momordica charantia* L. "balsamino" | F |
| *Sabicea villosa* H.B.K. | F |
| *Aegiphila panamensis* Moldenke | O |
| *Chamissoa altissima* (Jacq.) H.B.K. | O |
| *Cissampelos pareira* L. | O |
| *Cissampelos tropaeolifolia* DC. | O |
| *Eupatorium sinclairii* Benth. | O |
| *Gurania seemanniana* Cogn. "ya te vi" | O |
| *Stigmatophyllon ellipticum* (H.B.K.) Juss. | O |
| *Stigmatophyllon lindenianum* Juss. | O |
| *Tournefortia angustifolia* R. and P. | O |
| *Alternanthera williamsii* (Standl.) Standl. | R |
| Epiphytes | |
| *Codonanthe crassifolia* (Focke) Morton | O |
| *Browneopsis excelsa* Pittier "cuchillito" | D |
| *Erythrina glauca* Willd. "gallito" | D |
| *Pithecellobium longifolium* (H. and B.) Standl. "pichinde" | D |
| *Raphia taedigera* Mart. "pangana" (Atlantic side only) | D |
| *Astrocaryum standleyanum* Bailey "chunga" | F |
| *Cecropia longipes* Pitt. "guarumo" | F |
| *Cecropia obtusifolia* Bertol. "guarumo" | F |
| *Cecropia peltata* L. "guarumo" | F |
| *Chomelia spinosa* Jacq. | F |
| *Clusia spp.* "copey" | F |
| *Geoffroya inermis* W. Wright "cocu" | F |
| *Gustavia superba* (H.B.K.) Berg. "membrillo" | F |

Appendix table 1.   List of species in Riverine forest, *cont.*

| Compartments | Relative abundance |
|---|:---:|
| *Inga punctata* Willd. "guava" | F |
| *Luehea seemanii* Tr. and Pl. "guacimo" | F |
| *Mouriri parvifolia* Benth. | F |
| *Myrcia splendens* (Sw.) DC. "camaroncito" | F |
| *Olmedia aspera* R. and P. | F |
| *Pentaclethra macroloba* (Willd.) Ktze. "gallinazo" | F |
| *Piper lucigaudens* DC. | F |
| *Pithecellobium rufescens* Benth. | F |
| *Posoqueria latifolia* (Rudge) R. and S. "boca vieja" | F |
| *Quararibea asterolepis* Pitt. "punula" | F |
| *Quararibea bracteolosa* (Ducke) Cuatr. | F |
| Subcanopy trees | |
| *Quararibea pterocalyx* Hemsl. | F |
| *Trichanthera gigantea* (H.B.K.) H. and B. "asedera" | F |
| *Warszewiczia coccinea* (Vahl) Klotzsh | F |
| *Annona glabra* L. "anon de puerco" | O |
| *Byrsonima crassifolia* (L.) H.B.K. "nance" | O |
| *Capparis flexuosa* L. | O |
| *Carapa guianensis* Aubl. "tangaré" | O |
| *Casearia nitida* (L.) Jacq. "raspa lengua" | O |
| *Cupania costaricensis* Radlk. | O |
| *Diphysa robinioides* Benth. "macano" | O |
| *Ficus hartwegii* (Miq.) Miq. | O |
| *Ficus maxima* Mill. | O |
| *Ficus nymphaeifolia* L. | O |
| *Ficus perforata* L. "mata palo" | O |
| *Genipa americana* L. "jagua" | O |
| *Helicteres guazumaefolia* H.B.K. | O |
| *Inga coruscans* Willd. | O |
| *Inga pauciflora* Walp. & Duchass. | O |
| *Inga spectabilis* (Vahl) Willd. "guava" | O |
| *Lonchocarpus sericeus* (Poir.) DC. | O |
| *Mangifera indica* L. "mango" | O |
| *Margaritaria nobilis* L.f. | O |
| *Rinorea squamata* Blake | O |
| *Sapindus saponaria* L. "jaboncillo" | O |
| *Spondias mombin* L. "jobo" | O |
| *Trichilia cipo* (A. Juss.) C. DC. | O |
| *Urera baccifera* (L.) Gaud. "ortiga" | O |
| *Urera caracasana* (Jacq.) Griseb. | O |
| *Gliricidia sepium* (Jacq.) Steud. "mata ratón" | R |
| *Piper taboganum* C. DC. | R |
| *Scheelia zonensis* Bailey "palma real" | R |

Appendix table 1.    List of species in Riverine forest, *cont.*

| Compartments | Relative abundance |
|---|---|
| Canopy trees | |
| *Pachira aquatica* Aubl. "sapote longo" | D |
| *Prioria copaifera* Griseb. "cativo" | D |
| *Pterocarpus hayesii* Hemsl. "sangre de gallo" | D |
| *Pterocarpus officinalis* Jacq. "suela" | D |
| *Swartzia panamensis* Benth. "cutarro" | D |
| *Tabebuia pentaphylla* (L.) Hemsl. "roble" | D |
| *Copaifera aromatica* Dwyer "cabismo" | F |
| *Ficus insipida* Willd. "higuerón" | F |
| *Anacardium excelsum* (B. and B.) Skeels "espavé" | O |
| *Carapa nicaraguensis* C. DC. | O |
| *Platymiscium polystachyum* Benth. "quirá" | O |
| *Pseudobombax septenatum* (Jacq.) Dug. "barrigón" | O |
| *Hura crepitans* L. "nuno" | R |

Appendix table 2.  List of species in Mangrove forest.

| Compartments | Relative abundance |
|---|---|
| Understory herbs | |
| *Crinum darienensis* Woodson | O |
| *Sesuvium portulacastrum* L. "verdolaga" | O |
| *Pavonia rhizophorae* Killip | R |
| Understory palms | |
| *Bactris major* Jacq. "mongo lolo" | O |
| Understory shrubs | |
| *Conocarpus erecta* L. "mangle boton" | F |
| *Tabebuia palustris* Hemsl. | F |
| *Conostegia subcrustulata* (Beurl.) Triana | O |
| Woody vines | |
| *Omphalea diandra* L. | O |
| *Pithecellobium hymeneaefolium* (H. & B.) Benth. | O |
| *Rhabdadenia paludosa* (Vahl) Miers | O |
| *Topobea membranacea* Wurdack | O |

Appendix table 2.    List of species in Mangrove forest, *cont.*

| Compartments | Relative abundance |
|---|---|
| Canopy trees | |
| *Avicennia bicolor* Standl. "mangle negro" | D |
| *Avicennia germinans* (L.) L. "mangle negro" | D |
| *Laguncularia racemosa* (L.) Gaert. "mangle blanco" | D |
| *Pelliciera rhizophorae* Tr. and Pl. "mangle piñuela" | D |
| *Rhizophora spp.* "mangle colorado" | D |
| *Citharexylum caudatum* L. | O |
| *Hibiscus tiliaceus* L. "majagua" | O |
| *Mora oleifera* (Tr.) Ducke "alcornoque" | R |

Appendix table 3.    List of species in Premontane Wet forest.

| Compartments | Relative abundance |
|---|---|
| Understory herbs | |
| *Coccocypselum herbaceum* Lam. | F |
| *Spigelia anthelmia* L. | F |
| *Aciotis rostellata* (Naud.) Triana | O |
| *Begonia garagarana* C. DC. | O |
| *B. nelumbiifolia* Schlecht. and Cham. | O |
| *Campelia zanonia* (L.) H.B.K. | O |
| *Chrysothemis friedrichsthaliana* (Hanst.) Moore | O |
| *Cleome panamensis* Standl. | O |
| *Cyathula achyranthoides* (H.B.K.) Moq. | O |
| *Dorstenia contrayerba* L. | O |
| *Episcia lilacina* Hanst. | O |
| *Erechtites hieracifolia* (L.) Raf. | O |
| *Geophila herbacea* (L.) Schum. | O |
| *Monolaena ovata* Cogn. | O |
| *(Nautilocalyx* sp. nov.) | O |
| *Pavonia fruticosa* (Mill.) Fawcett and Rendle | O |
| *P. rosea* Schlecht. | O |
| *Peperomia urocarpoides* C. DC. | O |
| *Pilea ptericlada* Donn. Sm. | O |
| *Triolena hirsuta* (Benth.) Triana | O |
| *Pfaffia grandiflora* (Hook.) R.E.Fr. | R |
| *Pilea pubescens* Liebm. | R |

Appendix table 3.   List of species in Premontane Wet forest, *cont.*

| Compartments | Relative abundance |
|---|:---:|
| Understory Palmitas, etc. | |
| *Costus nutans* K. Sch. | F |
| *Heliconia psittacorum* L.f. | F |
| *Renealmia cernua* (Sw.) MacBride | F |
| *Stromanthe lutea* (Jacq.) Eichl. | F |
| *Xiphidium caeruleum* Aubl. "palmita" | F |
| *Aechmea tillandsioides* (Mart.) Baker | O |
| *Bromelia karatas* L. "pita" | O |
| *B. pinguin* L. "piñuela" | O |
| *Carludovica palmata* R. and P. "portorico" | O |
| *Chamaedorea allenii* L. H. Bailey | O |
| *Costus villosissimus* Jacq. "cañagria" | O |
| *Cyclanthus bipartitus* Poit. "portorico" | O |
| *Dieffenbachia oerstedii* Schott | O |
| *D. pittieri* Engl. and Krause | O |
| *Geonoma congesta* H. Wendl. | O |
| *G. deversa* (Poit.) Kunth. | O |
| *G. obovata* Wendl. ex Spruce | O |
| *Heliconia rostrata* R. and P. | O |
| *Pitcairnia carnea* Beer | O |
| *Renealmia aromatica* (Aubl.) Griseb. | O |
| *R. exaltata* L.f. | O |
| *Spathiphyllum floribundum* (L. and A.) N.E. Br. | O |
| *S. friedrichsthalii* Schott | O |
| *Synechanthus warscewiczianus* H. Wendl. | O |
| *Thalia geniculata* L. | O |
| *Xanthosoma helleborifolium* (Jacq.) Schott | O |
| *Anthurium garagaranum* Standl. | R |
| *Carludovica killipii* Standl. | R |
| *C. sarmentosa* Sagot ex Drude | R |
| *C. utilis* (Oerst.) Benth. and Hook. | R |
| *Chamaedorea pygmaea* Wendl. | R |
| *C. terryorum* Standl. | R |
| *Phytelephas seemannii* Cook "tagua" | R |
| *Renealmia arundinaria* Woods. | R |
| *Rhodospatha forgeti* N.E. Br. | R |
| Understory shrubs | |
| *Mabea occidentalis* Benth. | D |
| *Aphelandra sinclairiana* Nees | F |
| *Clavija mezii* Pittier | F |
| *Clidemia dentata* Don | F |
| *C. gracilis* Pitt. | F |
| *C. ombrophylla* Gleason | F |

Appendix table 3.    List of species in Premontane Wet forest, *cont.*

| Compartments | Relative abundance |
|---|:---:|
| *C. purpureo-violacea* Cogn. | F |
| *C. reitziana* Cogn. and Gl. | F |
| *C. septuplinervia* Cogn. | F |
| *C. tococoidea* (DC.) Gl. | F |
| *Coccoloba coronata* Jacq. | F |
| *C. manzanillensis* Beurl. "hueso" | F |
| *Conostegia bracteata* Tr. | F |
| *C. micromeris* Standl. | F |
| *C. setosa* (Tr.) Gl. | F |
| *C. subcrustulata* (Beurl.) Triana | F |
| *Faramea luteovirens* Standl. | F |
| *F. occidentalis* (L.) Rich. "huesito" | F |
| *Hasseltia floribunda* H.B.K. "parimontón" | F |
| *Henrietella fascicularis* (Sw.) Tr. | F |
| *Herrania purpurea* (Pitt.) R.E. Schultes "cacao cimarron" | F |
| *Leandra dichotoma* (Don) Cogn. | F |
| *Miconia barbinervius* (Benth.) Tr. | F |
| *M. centrodesma* Naud. | F |
| *M. disparilis* (Standl.) L.O. Wms. | F |
| *M. goniostigma* Tr. | F |
| *M. gracilis* Tr. | F |
| *M. micromeris* Standl. | F |
| *M. nervosa* (Sm.) Tr. | F |
| *M. nigricans* Cogn. | F |
| *M. oinochrophylla* Donn. Sm. | F |
| *Palicourea guianensis* Aubl. | F |
| *Picramnia antidesma* Sw. | F |
| *Pothomorphe peltata* (L.) Miq. "hinojo" | F |
| *Psychotria involucrata* Sw. | F |
| *Quassia amara* L. "guavito amargo" | F |
| *Aphelandra deppeana* Cham. and Schl. | O |
| *Appunia* sp. | O |
| *Bauhinia glabra* Jacq. | O |
| *Bertiera guianensis* Aubl. | O |
| *Cephaelis ipecacahuana* (Brot.) Rich. "raicilla" | O |
| *C. tomentosa* (Aubl.) Vahl | O |
| *Clidemia capitellata* (Bonpl.) D. Don | O |
| *Cordia bicolor* DC. | O |
| *C. ferruginea* (Lam.) R. and S. | O |
| *Gonzalagunia rosea* Standl. | O |
| *Hamelia axillaris* Sw. | O |
| *H. pauciflora* Standl. | O |
| (*Hansteinia* sp. nov.) | O |

Appendix table 3.    List of species in Premontane Wet forest, *cont.*

| Compartments | Relative abundance |
|---|---|
| *Hiraea obovata* Ndzu. | O |
| *Hoffmannia ghiesbrechtii* Hemsl. | O |
| *Hybanthus prunifolius* (Schult.) Schulze | O |
| *Isertia haenkeana* DC. | O |
| *Machaonia acuminata* H.B.K. "espino blanco" | O |
| *Malvaviscus arboreus* Cav. "papita de monte" | O |
| *Miconia splendens* (Sw.) Griseb. | O |
| *Myriocarpa yzabalensis* (Donn. Sm.) Killip | O |
| *Ossaea brenesii* Standl. | O |
| *O. diversifolia* (Bonpl.) Cogn. | O |
| *Palicourea triphylla* DC. | O |
| *P. galleottiana* Martens | O |
| Understory shrubs | |
| *Piper acutissimum* Trel. | O |
| *P. aequale* Vahl | O |
| *P. carrilloanum* C. DC. | O |
| *P. grande* Vahl | O |
| *P. latibracteatum* C. DC. | O |
| *P. pseudogaragaranum* Trel. | O |
| *P. sperdinum* C. DC. | O |
| *P. subcaudatum* Trel. | O |
| *Pleuropetalum pleiogynum* (O. Ktze.) Standl. | O |
| *Psychotria alboviridula* Krause | O |
| *P. brachiata* Sw. | O |
| *P. chagrensis* Standl. | O |
| *P. cuspidata* Bredem. | O |
| *P. emetica* L.f. "raicillo macho" | O |
| *P. graciliflora* Benth. | O |
| *P. grandis* Sw. | O |
| *P. horizontalis* Sw. | O |
| *P. luxurians* Rusby | O |
| *P. marginata* Sw. | O |
| *P. uliginosa* Sw. | O |
| *Rondeletia panamensis* DC. "candelo" | O |
| *Roupala montana* Aubl. | O |
| *Solanum diphyllum* L. | O |
| *S. hirtum* Vahl | O |
| *S. rugosum* Dunal | O |
| *Tabernaemontana chrysocarpa* Blake | O |
| *Tocoa acuminata* Benth. | O |
| *Topobea pluvialis* Standl. | O |
| *Urera elata* (Sw.) Griseb. | O |
| *Biophytum* Sp | R |

Appendix table 3.    List of species in Premontane Wet forest, *cont.*

| Compartments | Relative abundance |
|---|:---:|
| *Clidemia* cf. *myrmecina* Gleason | R |
| *Henrietella fissanthera* Gleason | R |
| *Leandra subultata* Gleason | R |
| *Maytenus* sp. | R |
| *Miconia calvescens* DC. | R |
| *Mollinedia costaricensis* Donn. Sm. | R |
| *M. darienensis* Standl. | R |
| *Piper lucigaudens* DC. | R |
| *Siparuna diandra* J. Duke | R |
| Woody vines | |
| *Drymonia spectabilis* (H.B.K.) Mart | F |
| *Passiflora vitifolia* H.B.K. "granadillo de monte" | F |
| *Adelobotrys adscendens* (Sw.) Triana | O |
| *Arrabidaea panamensis* Sprague | O |
| (*Blakea hirsuta*?) | O |
| *Byttneria aculeata* Jacq. "espino hueco" | O |
| *Columnea flavida* Morton | O |
| *Corynostylis arborea* (L.) Blake | O |
| *Cyphomandra costaricensis* D. Sm. "contra gallinazo" | O |
| *Doliocarpus dentatus* (Aubl.) Standl. | O |
| *Hiraea fagifolia* (DC.) Juss. | O |
| *Hylenaea praecelsa* (Miers) A. C. Smith "colmillo de puerco" | O |
| *Machaerium arboreum* (Jacq.) Vogel | O |
| *Maripa panamensis* Hemsl. | O |
| *Martinella obovata* (H.B.K.) Bur. & Sch. | O |
| *Petrea aspera* Turcz. "flor de niña" | O |
| *Phryganocydia corymbosa* (Vent.) Bur. | O |
| *Pithecellobium hymenaeafolium* (H. and B.) Benth. | O |
| *Strychnos darienensis* Seem. | O |
| *Strychnos panamensis* Seem. "canjura" | O |
| *Topobea membranacea* Wurdack | O |
| *Topobea urophylla* Standl. | O |
| *Topobea watsonii* Cogn. | O |
| *Trichostigma octandrum* (L.) H. Walt. | O |
| *Vitis tiliaefolia* H. and B. "uva" | O |
| *Wulffia baccata* (L.f.) Kuntze | O |
| *Aristolochia sylvicola* Standl. | R |
| *Cheiloclinum cognatum* (Miers) A.C. Sm. | R |
| *Piper grande* Vahl | R |
| *Rourea pittieri* S.F. Blake | R |
| Herbaceous vines | |
| *Clidemia epiphytica* (Tr.) Cogn. | O |
| *Dalechampia tiliaefolia* Lam. | O |

Appendix table 3.   List of species in Premontane Wet forest, *cont.*

| Compartments | Relative abundance |
|---|---|
| *Eupatorium iresineoides* H.B.K. | O |
| *Eupatorium macrophyllum* L. | O |
| *Eupatorium microstemon* Cass. | O |
| *Eupatorium odoratum* L. | O |
| *Manettia coccinea* (Aubl.) Willd. | O |
| *Mikania micrantha* H.B.K. | O |
| *Stigmatophyllon lindenianum* Juss. | O |
| *Chamissoa maximiliana* Mart. | R |
| *Clidemia oblonga* Gl. | R |
| Epiphytes | |
| *Aechmea pubescens* Baker | O |
| *Aechmea tillandsioides* (Mart.) Baker | O |
| *Anthurium acutangulum* Engl. | O |
| *Anthurium crassinervium* (Jacq.) Schott | O |
| *Anthurium gracile* (Rudge) Lindl. | O |
| *Anthurium pittieri* Engl. | O |
| *Anthurium scandens* (Aubl.) Engl. | O |
| *Anthurium scolopendrinum* (Ham.) Kunth. | O |
| *Anthurium williamsii* Krause | O |
| *Begonia guaduensis* H.B.K. | O |
| *Codonanthe crassifolia* (Focke) Morton | O |
| *Columnea consanguinea* Beurl. | O |
| *Drymonia serrulata* (Jacq.) Mart. ex DC. | O |
| *Epidendrum crassilabium* Poepp. and Endl. | O |
| *Epiphyllum phyllanthus* (L.) Haw. | O |
| *Guzmania coriostachya* (Griseb.) Mez | O |
| *Guzmania glomerata* Mez and Werckle | O |
| *Guzmania minor* Mez | O |
| *Monstera adansonii* Schott | O |
| *Philodendron tripartitum* (Jacq.) Schott | O |
| *Phoradendron piperoides* (H.B.K.) Nutt. | O |
| *Phoradendron supravenulosum* Trel. "mata palo" | O |
| *Sphyrospermum buxifolium* P. and E. | O |
| *Stenospermation spruceanum* Schott | O |
| *Tillandsia crispa* (Baker) Mez | O |
| *Tillandsia monadelpha* (E. Morr.) Baker | O |
| *Topobea praecox* Gleason | O |
| Subcanopy trees | |
| *Oenocarpus panamanus* Bailey "maquenque" | D |
| *Brownea macrophylla* Lind. | F |
| *Brownea rosa-de-monte* Berg. "rosa-de-monte" | F |
| *Cespedesia macrophylla* Seem. "membrillo" | F |
| *Chomelia spinosa* Jacq. | F |

Appendix table 3.    List of species in Premontane Wet forest, *cont.*

| Compartments | Relative abundance |
|---|:---:|
| *Clusia spp.* "copey" | F |
| *Conostegia xalapensis* (Bonpl.) D. Don | F |
| *Gustavia superba* (H.B.K.) Berg. "membrillo" | F |
| *Heisteria longipes* Standl. | F |
| *Iriartea corneto* (Karst.) Wendl. "jira" | F |
| *Luehea seemannii* Tr. and Pl. "guacimo" | F |
| *Miconia borealis* Gleason | F |
| *Ouratea lucens* (H.B.K.) Engl. | F |
| *Pentagonia brachyotis* Standl. | F |
| *Pentagonia macrophylla* Benth. "hoja de murcielago" | F |
| *Pogonopus speciosus* (Jacq.) Schum. | F |
| *Posoqueria latifolia* (Rudge) R. and S. "boca vieja" | F |
| *Pourouma scobina* R. Ben. "guarumo macho" | F |
| *Siparuna pauciflora* (Beurl.) DC. "pasmo" "cuama" | F |
| *Stemmadenia grandiflora* (Jacq.) Woodson | F |
| *Warszewiczia coccinea* (Vahl) Klotzsh | F |
| *Xylopia frutescens* Aubl. "malagüeta" | F |
| *Ahovaia nitida* Pichon | O |
| *Albizia carbonaria* Britton | O |
| *Apeiba membranacea* Spruce ex Benth. | O |
| *Apeiba tibourbou* Aubl. "peine de mico" | O |
| *Banara guianensis* Aubl. | O |
| *Bauhinia ligulata* Pitt. | O |
| *Bellucia costaricensis* Cogn. | O |
| (*Borojoa sp. nov.*) "borojó" | O |
| *Bunchosia cornifolia* H.B.K. | O |
| *Byrsonima coriacea* (Sw.) DC. "nancillo" | O |
| *Calycophyllum candidissimum* (Vahl.) DC. "madroño" | O |
| *Caraipa sp.* | O |
| *Casearia banquitans* Krause | O |
| *Casearia javitensis* H.B.K. | O |
| *Cassipourea elliptica* (Sw.) Poir. | O |
| *Castilla tunu* Hemsl. "caucho macho" | O |
| *Cecropia longipes* Pitt. "guarumo" | O |
| *Cecropia peltata* L. "guarumo" | O |
| *Ceiba rosea* (Seem.) K. Schum. | O |
| *Cestrum latifolium* Lam. "Juan de la verdad" | O |
| *Chione chambersii* Dwyer and Hayden | O |
| *Chomelia recordii* Standl. | O |
| *Chrysophyllum cainito* L. "caimito" | O |
| *Citharexylum caudatum* L. | O |
| *Coccoloba acuminata* H.B.K. | O |
| *Coccoloba coronata* Jacq. | O |

Appendix table 3.    List of species in Premontane Wet forest, *cont.*

| Compartments | Relative abundance |
|---|---|
| *Compsoneura sprucei* (A. DC.) Warb. | O |
| *Condaminea corymbosa* (R. and P.) DC. | O |
| *Conostegia puberula* Cogn. | O |
| *Cordia glabra* L. | O |
| *Couepia panamensis* Standl. | O |
| *Coussarea* sp. | O |
| *Cryosophila sp.* "nupa" | O |
| *Cryosophila albida* Bartl. | O |
| *Cupania fulvida* Tr. and Pl. "gorgojo" | O |
| *Cymbopetalum brasiliensis* (Vell.) Benth. | O |
| *Dendropanax arboreus* (L.) Desne. and Pl. "vaquero" | O |
| *Dialyanthera otoba* (H. and B.) Warb. "otoba" | O |
| *Dichapetalum axillare* Woodson | O |
| *Duguetia panamensis* Standl. | O |
| *Duguetia vallicola* Macbr. | O |
| *Ficus padifolia* H. and B. | O |
| *Ficus perforata* L. "mata palo" | O |
| *Ficus pertusa* L.f. | O |
| *Ficus trigonata* L. | O |
| *Fissicalyx fendleri* Benth. | O |
| *Genipa americana* L. "jagua" | O |
| *Geoffroya inermis* W. Wright "cocu" | O |
| *Gloeospermum portobellensis* A. Robyns | O |
| *Grias dukei* Dwyer | O |
| *Guapira standleyiana* Woodson | O |
| *Guarea kunthiana* A. Juss. | O |
| *Guarea tonduziana* C. DC. | O |
| *Guatteria* cf. *chiriquensis* R.E. Fr. | O |
| *Guatteria dumetorum* R.E. Fr. | O |
| *Guatteria panamensis* R.E. Fr. | O |
| *Guettarda macrosperma* Donn. Sm. | O |
| *Hampea appendiculata* (J.D. Sm.) Standl. | O |
| *Heisteria costaricensis* Donn. Sm. | O |
| *Henrietella tuberculosa* Donn. Sm. | O |
| *Hirtella triandra* Sw. "garrapato" | O |
| *Hyeronima laxifolia* (Tul.) Muell.-Arg. | O |
| *Inga caldasiana* Britt. and Killip | O |
| *Inga coruscans* Willd. | O |
| *Inga punctata* Willd. "guava" | O |
| *Inga quaternata* Poeppig | O |
| *Inga saffordiana* Pitt. | O |
| *Isertia hypoleuca* | O |
| *Lindackeria laurina* Presl "choco cucullo" | O |

Appendix table 3.    List of species in Premontane Wet forest, *cont.*

| Compartments | Relative abundance |
|---|:---:|
| (*Maba darienensis* Dwyer) | O |
| *Macrolobium pittieri* (Rose) Schery | O |
| *Malmea depressa* (Baill.) R.E. Fr. | O |
| *Manilkara sapotilla* (Jacq.) Gilly | O |
| *Margaritaria nobilis* L.f. | O |
| *Mayna longicuspis* Standl. | O |
| *Miconia argentea* (Sw.) DC. | O |
| *Miconia calvescens* DC. | O |
| *Miconia hondurensis* Donn. Sm. | O |
| *Miconia hyperprasina* Naud. | O |
| *Miconia microcarpa* DC. | O |
| *Miconia prasina* (Sw.) DC. | O |
| *Ochroma pyramidale* (Cav.) Urb. "balsa" | O |
| *Olmedia aspera* R. and P. | O |
| *Ossaea trichocalyx* Pitt. | O |
| *Ouratea flexipedicellata* Dwyer | O |
| *Ouratea patelliformis* Dwyer | O |
| *Patrisia pyrifera* L. Rich. | O |
| *Peltogyne purpurea* Pittier "nazareno" | O |
| *Perebea sp.* | O |
| *Pereskia bleo* (H.B.K.) DC. "naju" | O |
| *Piper imperiale* (Miq.) C. DC. | O |
| *Pithecellobium rufescens* Benth. | O |
| *Podocarpus sp.* "pino" | O |
| *Pothomorphe umbellata* (L.) Miq. | O |
| *Pourouma radula* R. Ben. | O |
| *Psychotria patens* Sw. "garricillo" | O |
| *Quararibea cordata* (H. and B.) Vischer "sapote" | O |
| *Randia armata* (Sw.) DC. "jagua macho" "miel quema" | O |
| *Rheedia madruno* (H.B.K.) Pl. and Tr. "madroño" | O |
| *Rinorea squamata* Blake | O |
| *Rollinia* cf. *pittieri* Saff. | O |
| *Sabal allenii* Bailey "guágara" | O |
| *Sapindus saponaria* L. "jaboncillo" | O |
| *Sapium biglandulosum* Muell.-Arg. "higo" | O |
| *Saurauia laevigata* Tr. and Pl. | O |
| *Scheelia zonensis* Bailey "palma real" | O |
| *Simaba cedron* Planch. "cedrón" | O |
| *Sterculia costaricana* Pitt. | O |
| *Talisia nervosa* Radlk. "mamón de monte" | O |
| *Ternstroemia tepazapote* Schlecht. and Cham. "manglillo" | O |
| *Theobroma bicolor* H. and B. "bacao" | O |
| *Trichilia hirta* L. | O |

Appendix table 3.    List of species in Premontane Wet forest, *cont.*

| Compartments | Relative abundance |
|---|---|
| *Trichilia montana* H.B.K. | O |
| *Trophis racemosa* (L.) Urb. "ojoche macho" | O |
| *Urera baccifera* (L.) Gaud. "ortiga" | O |
| *Urera caracasana* (Jacq.) Grisebl | O |
| *Virola koschnyi* Warb. | O |
| *Virola sebifera* Aubl. "fruta dorado" | O |
| *Albizia caribaea* (Urban) Britt. and Rose | U |
| *Eschweilera panamensis* Pitt. | U |
| *Eschweilera pittieri* R. Kunth | U |
| *Lecythis ampla* Miers "coco" | U |
| *Nectandra gentlei* Lundell | U |
| Canopy trees | |
| *Anacardium excelsum* (B. and B.) Skeels "espavé" | D |
| *Bombacopsis quinata* (Jacq.) Dug. "cedro espinoso" | F |
| *Bombacopsis sessilis* (Benth.) Pitt. "ceibo" | F |
| *Brosimum guianense* (Aubl.) Huber | F |
| *Ceiba pentandra* (L.) Gaertn. "bongo" | F |
| (*Cochlospermum n. sp.* Robyns) "poroporo" | F |
| *Myroxylon balsamum* (L.) Harms "bálsamo" | F |
| *Oleiocarpon panamense* (Pittier) Dwyer "almendro" | F |
| (*Aspidosperma dukei* Dwyer) | O |
| *Brosimum utile* (H.B.K.) Pitt. "palo de vaca" | O |
| *Enterolobium cyclocarpum* (Jacq.) Griseb. "corotú" | O |
| *Ficus insipida* Willd. "higuerón" | O |
| *Hymenaea courbaril* L. "algarroba" | O |
| *Jacaranda* cf. *caucana* Pittier "siete cueros" | O |
| *Jacaranda rhombifolia* G.F.W. Mey. "palo de buba" | O |
| *Lecythis tuyrana* Pittier "coco" | O |
| *Licania hypoleuca* Benth. | O |
| *Licania platypus* (Hemsl.) Fritsch? "tuqueso" | O |
| *Manilkara chicle* "chicle" | O |
| *Mimusops darienesis* Pitt. "níspero" | O |
| *Platypodium elegans* Vogel "canalua" | O |
| *Poeppigia procera* Presl | O |
| *Poulsenia armata* (Miq.) Standl. "cucua" | O |
| *Prioria copaifera* Griseb. "cativo" | O |
| *Swartzia panamensis* Benth. "cutarro" | O |
| *Tabebuia guayacan* (Seem.) Hemsl. "guayacán" | O |
| *Cariniana sp.* "abarco" | R |
| *Couroupita magnifica* Dwyer | R |
| *Dialyanthera latialata* Pittier | R |
| *Lecythis elata* Dwyer "coco" | R |
| *Platymiscium dariense* Dwyer | R |

Appendix table 3.    List of species in Premontane Wet forest, *cont.*

| Compartments | Relative abundance |
|---|---|
| *Pouteria izabalensis* (Standl.) Baehni | R |
| *Pouteria neglecta* Cronquist "zapote del monte" | R |
| *Pterocarpus officinalis* Jacq. "suela" | R |

Appendix table 4.    List of Species in Tropical Moist forest.

| Compartments | Relative abundance |
|---|---|
| Understory herbs | |
| *Aneilema geniculatum* (Jacq.) Woodson | O |
| *Begonia filipes* Benth. | O |
| *Chrysothemis friedrichstaliana* (Hanst.) Moore | O |
| *Commelina diffusa* Burm. f. | O |
| *Cyperus diffusus* Vahl | O |
| *Cyperus ferax* L. C. Rich. | O |
| *Dichorisandra hexandra* (Aubl.) Standl. | O |
| *Elephantopus mollis* H.B.K. | O |
| *Eleutheranthera ruderalis* (Sw.) Sch. Bip. | O |
| *Paspalum paniculatum* L. | O |
| *Paspalum virgatum* L. | O |
| *Pavonia fruticosa* (Mill.) Fawcett and Rendle | O |
| *Pavonia longipes* Standl. | O |
| *Pavonia paniculata* Cav. "pape" | O |
| *Pavonia rosea* Schlecht. | O |
| *Peperomia brevipeduncula* Trel. | O |
| *Peperomia cyclophylla* Miq. | O |
| *Peperomia quadrangularis* (Thomps.) A. Dietr. | O |
| *Peperomia rotundifolia* (L.) H.B.K. | O |
| *Petiveria alliacea* L. "anamú" | O |
| *Rhynchospora cephalotes* (L.) Vahl | O |
| *Scleria melaleuca* Uittien | O |
| *Scleria pterota* Presl "cortadero" | O |
| *Scleria setulosa-ciliata* Boeckl. | O |

Appendix table 4.  List of species in Tropical Moist forest, *cont.*

| Compartments | Relative abundance |
|---|:---:|
| *Spigelia anthelmia* L. | O |
| *Triolena hirsuta* (Benth.) Triana | O |
| *Cleome pubescens* Sims | R |
| *Cleome serrata* Jacq. | R |
| *Cyperus esculentus* L. | R |
| *Echinodorus longipetalus* Micheli | R |
| *Episcia lilacina* Hanst. | R |
| *Fimbristylis miliacea* (L.) Vahl | R |
| *Sida glomerata* Cav. "escoba" | R |
| *Sida pyramidata* Desf. | R |
| *Sida rhombifolia* L. | R |
| Understory Palmitas, etc. | |
| *Heliconia psittacorum* L.f. | F |
| *Heliconia subulata* R. and P. | F |
| *Renealmia cernua* (Sw.) MacBride | F |
| *Xiphidium caeruleum* Aubl. "palmita" | F |
| *Aechmea magdalenae* (Andre) Andre ex Baker "pita" | O |
| *Bromelia karatas* L. "pita" | O |
| *Bromelia pinguin* L. "pinuela" | O |
| *Calathea allouia* (Aubl.) Lindl. "bijao" | O |
| *Calathea altissima* (P. and E.) Koern. | O |
| *Calathea insignis* Peters | O |
| *Calathea lutea* (Aubl.) G.F.W. Meyer "hoja blanca" | O |
| *Carludovica palmata* R. and P. "portorico" | O |
| *Carludovica pittieri* Woodson | O |
| *Corozo oleifera* (H.B.K.) Bailey "corozo" | O |
| *Costus nutans* K. Sch. | O |
| *Costus ruber* Griseb. | O |
| *Costus villosissimus* Jacq. "cañagria" | O |
| *Dieffenbachia pittieri* Engl. and Krause | O |
| *Heliconia acuminata* L.C. Rich. | O |
| *Heliconia mariae* Hook. F. | O |
| *Heliconia platystachys* Baker | O |
| *Ischnosiphon leucophaeus* (P. and E.) Koern. | O |
| *Renealmia aromatica* (Aubl.) Griseb. | O |
| *Stromanthe lutea* (Jacq.) Eichl. | O |
| *Canna indica* L. | R |
| *Carludovica integrifolia* Woodson | R |
| *Cyclanthus bipartitus* Poit. "portorico" | R |
| *Phytelephas seemannii* Cook "tagua" | R |
| *Thalia geniculata* L. | R |
| *Zamia skinneri* Warscz. | R |

Appendix table 4.    List of species in Tropical Moist forest, *cont.*

| Compartments | Relative abundance |
|---|---|
| Understory shrubs | |
| *Faramea luteovirens* Standl. | D |
| *Mabea occidentalis* Benth. | D |
| *Piper pinoganense* Trel. | D |
| *Alibertia edulis* (L. Rich.) A. Rich. "trompito" | F |
| *Aphelandra sinclairiana* Nees | F |
| *Cephaelis tomentosa* (Aubl.) Vahl | F |
| *Clavija mezii* Pittier | F |
| *Hasseltia floribunda* H.B.K. "parimontón" | F |
| *Herrania purpurea* (Pitt.) R.E. Schultes "cacao cimarrón" | F |
| *Leandra dichotoma* (Don.) Cogn. | F |
| *Neea laetivirens* Standl. | F |
| *Palicourea guianensis* Aubl. | F |
| *Palicourea triphylla* DC. | F |
| *Piper darienense* C. DC. "duerme boca" | F |
| *Piper reticulatum* L. | F |
| *Quassia amara* L. "guavito amargo" | F |
| *Ahovaia nitida* Pichon | O |
| *Bauhinia glabra* Jacq. | O |
| *Bertiera guianensis* Aubl. | O |
| *Cassia fruticosa* Mill. | O |
| *Cassia reticulata* Willd. "laureño" | O |
| *Cephaelis elata* Sw. | O |
| *Chiococca alba* (L.) Hitchc. "lágrimas de Maria" | O |
| *Clibadium appressipilum* Blake | O |
| *Hybanthus prunifolius* (Schult.) Schulze | O |
| *Isertia hypoleuca* Benth. | O |
| *Leucaena multicapitulata* Schery | O |
| *Malvastrum americanum* (L.) Torr. | O |
| *Malvaviscus arboreus* Cav. "papito de monte" | O |
| *Miconia lacera* (Bonpl.) Naud. | O |
| *Miconia rufostellulata* Pitt. | O |
| *Miconia schlimii* Triana | O |
| *Neurolaena lobata* (L.) R.Br. "contragavilán" | O |
| *Ossaea diversifolia* (Bonpl.) Cogn. | O |
| *Piper aequale* Vahl | O |
| *Pouzolzia occidentalis* (Liebm.) Wedd. | O |
| *Prockia crucis* L. "huesito" | O |
| *Psychotria grandis* Sw. | O |
| *Psychotria horizontalis* Sw. | O |
| *Psychotria rufescens* H. and B. | O |
| *Rudgea cornifolia* (H. and B.) Standl. | O |
| *Siparuna guianensis* Aubl. "pasmo" | O |

Appendix table 4.  List of species in Tropical Moist forest, *cont.*

| Compartments | Relative abundance |
|---|:---:|
| *Solanum hirtum* Vahl | O |
| *Solanum rugosum* Dunal | O |
| *Steriphoma macranthum* Standl. | O |
| *Veronia canescens* H.B.K. "hierba de San Juan" | O |
| *Vernonia patens* H.B.K. "lengua de vaca" | O |
| *Acalypha macrostachya* Jacq. | R |
| *Callicarpa acuminata* H.B.K. | R |
| *Cephaelis ipecacahuana* (Brot.) Rich. "raicilla" | R |
| (*Hoffmannia darienensis sp.* nov.) | R |
| *Miconia impetiolaris* (Sw.) DC. | R |
| *Miconia oinochrophylla* Donn. Sm. | R |
| *Piper latibracteatum* C. DC. | R |
| *Siparuna nicaraguensis* Hemsl. "limoncillo" | R |
| *Tococa acuminata* Benth. | R |
| Woody vines | |
| *Connarus panamensis* Griseb. | F |
| *Connarus williamsii* Britt. | F |
| *Doliocarpus dentatus* (Aubl.) Standl. | F |
| *Doliocarpus major* Gmel. | F |
| *Machaerium capote* Tr. ex Dugand | F |
| *Maripa panamensis* Hemsl. | F |
| *Passiflora vitifolia* H.B.K. "granadillo de monte" | F |
| *Rourea glabra* H.B.K. | F |
| *Tetracera volubilis* L. "bejuco de agua" | F |
| *Wulffia baccata* (L.f.) Kuntze | F |
| *Acacia tenuifolia* (L.) Willd. | O |
| *Allamanda cathartica* L. | O |
| *Anthodon panamense* A.C. Smith "bejuco de estrella" | O |
| *Aristolochia arborescens* L. "Guaco" | O |
| *Bauhinia eucomosa* Blake | O |
| *Byttneria aculeata* Jacq. "espino hueco" | O |
| *Combretum fruticosum* (Loefl.) Stuntz | O |
| *Combretum sambuensis* Pitt. | O |
| *Combretum secundum* Jacq. | O |
| *Cuervea kappleriana* (Miq.) A.C. Sm. | O |
| *Cyphomandra costaricensis* D. Sm. "contra gallinazo" | O |
| *Davilla aspera* (Aubl.) Ben. "bejuco chumico" | O |
| *Desmoncus isthemensis* Bailey "matamba" | O |
| *Hippocratea volubilis* L. | O |
| *Machaerium longifolium* Benth. | O |
| *Martinella obovata* (H.B.K.) Bur. and Sch. | O |
| (*Mucuna bracteata* Dwyer) "ojo de venado" | O |
| *Omphalea diandra* L. | O |

Appendix table 4.  List of species in Tropical Moist forest, *cont.*

| Compartments | Relative abundance |
|---|---|
| *Paullinia pterocarpa* | O |
| *Peperomia caudulilimba* C. DC. | O |
| *Petrea aspera* Turcz. "flor de niña" | O |
| *Phryganocydia corymbosa* (Vent.) Bur. | O |
| *Piptocarpha chontalensis* Baker | O |
| *Prestonia portobellensis* (Beurl.) Woods. | O |
| *Serjania atrolineata* Sauv. and Wr. | O |
| *Smilax spinosa* Jacq. "zarza" | O |
| *Strychnos panamensis* Seem. "canjura" | O |
| *Tetracera portobellensis* Beurl. | O |
| *Aristolochia costaricensis* (Klotzsch) Duch. | R |
| *Aristolochia grandiflora* Sw. | R |
| *Aristolochia odoratissima* L. | R |
| *Aristolochia veraguensis* Duch. | R |
| Herbaceous vines | |
| *Philodendron guttiferum* Kunth. | F |
| *Bonamia maripoides* Hall. f. | O |
| *Centrosema plumieri* (Turp.) Benth. | O |
| *Clidemia epiphytica* (Tr.) Cogn. | O |
| *Desmodium canum* (Gmel.) "pegapega" | O |
| *Dioclea guianensis* Benth. "haba de monte" | O |
| *Eupatorium macrophyllum* L. | O |
| *Eupatorium microstemon* Cass. | O |
| *Eupatorium sinclairii* Benth. | O |
| *Gurania coccinea* Cogn. "ya te vi" | O |
| *Gurania seemanniana* Cogn. "ya te vi" | O |
| *Mendoncia lindavii* Rusby | O |
| *Mikania guaco* H.B.K. "guaco" | O |
| *Momordica charantia* L. "balsamino" | O |
| *Rhynchosia pyramidalis* (Lam.) Urb. "peronilla" | O |
| Epiphytes | |
| *Tillandsia kegeliana* Mez | F |
| *Aechmea tillandsioides* (Mart.) Baker | O |
| *Anthurium crassinervium* (Jacq.) Schott | O |
| *Anthurium holtonianum* Schott | O |
| *Anthurium myosuroides* (H.B.K.) Endl. | O |
| *Condonanthe crassifolia* (Focke) Morton | O |
| *Drymonia serrulata* (Jacq.) Mart. ex DC. | O |
| *Hylocereus monacanthus* (Lem.) Britt. and Rose | O |
| *Philodendron tripartitum* (Jacq.) Schott | O |
| *Topobea pluvialis* Standl. | O |
| *Wittia panemensis* Britt. and Rose "colacayman" | O |

Appendix table 4.    List of species in Tropical Moist forest, *cont.*

| Compartments | Relative abundance |
|---|---|
| Subcanopy trees | |
| *Mouriri parviflora* Benth. | D |
| *Allophylus occidentalis* (Sw.) Radlk. | F |
| *Alseis blackiana* Hemsl. | F |
| *Amaioua corymbosa* H.B.K. "madroño" | F |
| *Astrocaryum standleyanum* Bailey "chunga" | F |
| *Calycophyllum candidissimum* (Vahl) DC. "madroño" | F |
| *Chamaedorea* cf. *pacaya* Oerst. | F |
| *Chamaedorea wendlandiana* (Oerst.) Hemsl. | F |
| *Chomelia recordii* Standl. | F |
| *Chomelia spinosa* Jacq. | F |
| *Clusia spp.* "copey" | F |
| *Cnestidium rufescens* Planch. | F |
| *Cordia alliodora* (R. and P.) R. and S. "laurél" | F |
| *Guarea guidonia* (L.) Sleumer "cedro macho" | F |
| *Gustavia superba* (H.B.K.) Berg. "membrillo" | F |
| *Heisteria longipes* Standl. | F |
| *Hirtella racemosa* Lam. | F |
| *Luehea seemannii* Tr. and Pl. "guacimo" | F |
| *Miconia argentea* (Sw.) DC. | F |
| *Miconia borealis* Gleason | F |
| *Myrcia splendens* (Sw.) DC. "camaroncito" | F |
| *Neea amplifolia* Donn. Sm. | F |
| *Ouratea lucens* (H.B.K.) Engl. | F |
| *Piper pinoganense* Trel. | F |
| *Pithecellobium rufescens* Benth. | F |
| *Posoqueria latifolia* (Rudge) R. and S. "boca vieja" | F |
| *Protium panamensis* (Rose) Johnston "fruta de loro" | F |
| *Quararibea asterolepis* Pitt. "punula" | F |
| *Quararibea bracteolosa* (Ducke) Cuatr. | F |
| *Randia armata* (Sw.) DC. "jagua macho" "miel quema" | F |
| *Sabal allenii* Bailey "gúagara" | F |
| *Scheelia zonensis* Bailey "palma real" | F |
| *Siparuna pauciflora* (Beurl.) DC. "pasmo" "cuama" | F |
| *Sorocea affinis* Hemsl. "cauchillo" | F |
| *Stemmadenia grandiflora* (Jacq.) Woodson | F |
| *Swartzia simplex* (Sw.) Gaertn. "naranjita" | F |
| *Trophis racemosa* (L.) Urb. "ojoche macho" | F |
| *Xylopia aromatica* (Lam.) Eichl. | F |
| *Xylopia bocatorena* Schery | F |
| *Xylopia frutescens* Aubl. "malagüeta" | F |
| *Annona purpurea* Moc. and Sesse "guañabana toretc" | O |
| *Annona spraguei* Saff. "chirimoya de monte" | O |

Appendix table 4.   List of species in Tropical Moist forest, *cont.*

| Compartments | Relative abundance |
| --- | --- |
| *Apeiba membranacea* Spruce ex Benth. | O |
| *Apeiba tibourbou* Aubl. "peine de mico" | O |
| *Banara guianensis* Aubl. | O |
| *Bellucia costaricensis* Cogn. | O |
| *Brownea rosa-de-monte* Berg. "rosa-de-monte" | O |
| *Bunchosia cornifolia* H.B.K. | O |
| *Bursera simaruba* Sarg. "almácigo" | O |
| *Byrsonima coriacea* (Sw.) DC. "nancillo" | O |
| *Calophyllum longifolium* Willd. "maria" | O |
| *Capparis baducca* L. | O |
| *Capparis flexuosa* L. | O |
| *Carica papaya* L. "papaya" | O |
| *Casearia nitida* (L.) Jacq. "raspa lengua" | O |
| *Cassia leiophylla* Vogel | O |
| *Cassipourea elliptica* (Sw.) Poir. | O |
| *Castilla elastica* Cerv. "caucho" | O |
| *Castilla tunu* Hemsl. "caucho macho" | O |
| *Cecropia longipes* Pitt. "guarumo" | O |
| *Cecropia obtusifolia* Bertol. "guarumo" | O |
| *Cecropia peltata* L. "guarumo" | O |
| *Cedrela angustifolia* Sesse and Moc. "cedro" | O |
| *Ceiba rosea* (Seem.) K. Schum. | O |
| *Centrolobium paraense* Tul. | O |
| *Centrolobium yavizanum* Pittier "amarillo de Guayaquil" | O |
| *Cespedesia macrophylla* Seem. "membrillo" | O |
| *Cestrum latifolium* Lam. "Juan de la verdad" | O |
| *Chlorophora tinctoria* (L.) Gaud. "mora" | O |
| *Chomelia sp.* | *O* |
| *Chrysophyllum cainito* L. "caimito" | O |
| *Citharexylum caudatum* L. | O |
| *Conostegia xalapensis* (Bonpl.) D. Don | O |
| *Cordia glabra* L. | O |
| *Coussarea impetiolaris* Donn. Sm. "huesito" | O |
| *Cremastosperma anomalum* R. E. Fr. | O |
| *Croton panamensis* (Klotzsch) Muell-Arg. "sangrillo" | O |
| *Cupania costaricensis* Radlk. | O |
| *Cupania fulvida* Tr. and Pl. "gorgojo" | O |
| *Dendropanax arboreus* (L.) Desne. and Pl. "vaquero" | O |
| *Dialium guianensis* (Aubl.) Sandwith | O |
| *Didymopanax morototoni* (Aubl.) Dec. and Pl. "mangabé" | O |
| *Duguetia vallicola* Macbr. | O |
| *Eschweilera woodsoniana* Dwyer | O |
| *Ficus paraensis* (Miq.) Miq. | O |

Appendix table 4. List of species in Tropical Moist forest, *cont.*

| Compartments | Relative abundance |
|---|:---:|
| *Ficus nymphaeifolia* L. | O |
| *Ficus trigonata* L. | O |
| *Genipa americana* L. "jagua" | O |
| *Geoffroya inermis* W. Wright "cocu" | O |
| *Gloeospermum portobellensis* A. Robyns | O |
| *Grias dariensis* Dwyer | O |
| *Grias pittieri* R. Knuth. "jaguey" | O |
| *Guapira standleyiana* Woodson | O |
| *Guarea glabra* Vahl | O |
| *Guazuma ulmifolia* Lam. "guacimo" | O |
| *Guettarda odorata* (Jacq.) Lam. "huesito negro" | O |
| *Hirtella triandra* Sw. | O |
| *Inga mucuna* Walp. and Duchass. | O |
| *Inga punctata* Willd. "guava" | O |
| *Inga ruiziana* G. Don | O |
| *Inga saffordiana* Pitt. | O |
| *Inga spectabilis* (Vahl) Willd. "guava" | O |
| *Inga standleyiana* Pitt. | O |
| *Isertia hypoleuca* Benth. | O |
| *Lacistema aggregatum* (Berg.) Rusby | O |
| *Lonchocarpus sericeus* (Poir.) DC. | O |
| *Luehea speciosa* Willd. "guacimo" | O |
| *Machaerium chambersii* Dwyer | O |
| *Malpighia glabra* L. "grosella" | O |
| *Manilkara sapotilla* (Jacq.) Gilly | O |
| *Margaritaria nobilis* L.f. | O |
| *Mayna longicuspis* Standl. | O |
| *Miconia hondurensis* Donn. Sm. | O |
| *Miconia prasina* (Sw.) DC. | O |
| *Miconia splendens* (Sw.) Griseb. | O |
| *Mouriri completens* (Pitt.) Burret | O |
| *Ochroma pyramidale* (av.) Urb. "balsa" | O |
| *Oenocarpus panamanus* Bailey "maquenque" | O |
| *Olmedia aspera* R. and P. | O |
| *Peltogyne purpurea* Pittier "nazareno" | O |
| *Pentaclethra macroloba* (Willd.) Ktze. "gallinazo" | O |
| *Pentagonia brachyotis* Standl. | O |
| *Pentagonia macrophylla* Benth. "hoja de murciélago" | O |
| *Perebea sp.* | O |
| *Pereskia bleo* (H.B.K.) DC. "naju" | O |
| *Piper lucigaudens* DC. | O |
| *Piper sambuanum* C. DC. | O |
| *Piper subcaudatum* Trel. | O |

Appendix table 4.    List of species in Tropical Moist forest, *cont.*

| Compartments | Relative abundance |
|---|---|
| *Piper taboganum* C. DC. | O |
| *Piper villaramularum* C. DC. | O |
| *Pithecellobium longifolium* (H. and B) Standl. "pinchindé" | O |
| *Plumeria acutifolia* Poir. "caracucha" | O |
| *Pourouma scobina* R. Ben. "guarumo macho" | O |
| *Quararibea pterocalyx* Hemsl. | O |
| *Rauvolfia tetraphylla* L. | O |
| *Rheedia edulis* Tr. and Pl. "sastro" | O |
| *Rheedia madruno* (H.B.K.) Pl. and Tr. "madroño" | O |
| *Rinorea brachythrix* Blake | O |
| *Rhinorea sylvatica* (Seem.) O. Ktze. | O |
| *Rollinia chocoensis* R. E. Fr. | O |
| *Sapindus saponaria* L. "jaboncillo" | O |
| *Sapium biglandulosum* Muell.-Arg. "higo" | O |
| *Simaba cedron* Planch. "cedrón" | O |
| *Swartzia darienensis* Pitt. | O |
| *Talisia nervosa* Radlk. "mamón de monte" | O |
| *Tecoma stans* (L.) Juss. "copete" | O |
| *Trichanthera gigantea* (H.B.K.) H. and B. "nasedera" | O |
| *Trichilia cipo* (A. Juss.) C. DC. | O |
| *Trichilia hirta* L. | O |
| *Trichilia japurensis* C. DC. | O |
| *Trichilia montana* H.B.K. | O |
| *Trichospermum mexicanum* (DC.) Baill. | O |
| *Urera baccifera* (L.) Gaud. "ortiga" | O |
| *Virola sebifera* Aubl. "fruta dorado" | O |
| *Warszewiczia coccinea* (Vahl) Klotzsh | O |
| *Zanthoxylum panamensis* P. Wils. "arcabú" | O |
| *Zanthoxylum setulosum* P. Wils. "tachuelo" | O |
| Canopy trees | |
| *Anacardium excelsum* (B. & B.) Skeels "espavé" | D |
| *Cavanillesia platanifolia* (H. & B.) H.B.K. "cuipo" | D |
| *Ceiba pentandra* (L.) Gaertn. "bongo" | D |
| *Bombacopsis quinata* (Jacq.) Dug. "cedro espinoso" | F |
| *Bombacopsis sessilis* (Benth.) Pitt. "ceibo" | F |
| *Enterolobium cyclocarpum* (Jacq.) Griseb. "corotú" | F |
| *Licania hypoleuca* Benth. | F |
| *Platypodium elegans* Vogel "canalua" | F |
| *Pseudobombax septenatum* (Jacq.) Dug. "barrigón" | F |
| *Sterculia apetala* (Jacq.) Karst. "panamá" | F |
| *Terminalia amazonia* (J. F. Gmel.) Exell. | F |
| *Tetragastris panamensis* (Englm.) Ktze. "secuadro" | F |

Appendix table 4.    List of species in Tropical Moist forest, *cont.*

| Compartments | Relative abundance |
|---|---|
| *Vitex masoniana* Pitt. "cuajado" | F |
| (*Aspidosperma dukei* Dwyer) | O |
| *Brosimum guianense* (Aubl.) Huber | O |
| *Cassia grandis* L.f. | O |
| (*Cochlospermum sp. n.* Robyns) "poroporo" | O |
| *Copaifera aromatica* Dwyer "cabismo" | O |
| *Dipterodendron costaricensis* Radlk. "jarino" | O |
| *Ficus insipida* Willd. "higueron" | O |
| *Hirtella americana* L. | O |
| *Hymenaea courbaril* L. "algarroba" | O |
| *Jacaranda* cf. *caucana* Pittier "siete cueros" | O |
| *Jacaranda rhombifolia* G.F.W. Mey. "palo de buba" | O |
| *Lafoensia punicifolia* DC. "pino amarillo" | O |
| *Lecythis tuyrana* Pittier "coco" | O |
| *Licania platypus* (Hemsl.) Fritsch? "tuqueso" | O |
| *Manilkara chicle* (Pittier) Gilly "chicle" | O |
| *Mimusops darienensis* Pittier "nísper" | O |
| *Myroxylon balsamum* (L.) Harms "bálsamo" | O |
| *Oleiocarpon panamense* (Pittier) Dwyer "almendro" | O |
| *Platymiscium polystachyum* Benth. "quira" | O |
| *Poeppigia procera* Presl. | O |
| *Poulsenia armata* (Miq.) Standl. "cocua" | O |
| *Prioria copaifera* Griseb. "cativo" | O |
| *Protium sessiliflorum* (Rose) Standl. "sauco macho" | O |
| *Swartzia panamensis* Benth. "cutarro" | O |
| *Tabebuia guayacan* (Seem.) Hemsl. "guayacán" | O |
| *Tabebuia pentaphylla* (L.) Hemsl. "roble" | O |
| *Grias megacarpa* Dwyer | R |
| *Grias sternii* Dwyer "membrillo" | R |
| *Gyranthera dariensis* Pittier (San Blas) | R |
| *Hura crepitans* L. "nuno" | R |
| *Pachira aquatica* Aubl. "sapote longo" | R |
| *Pterocarpus hayesii* Hemsl. "sangre de gallo" | R |
| *Pterocarpus officinalis* Jacq. "suela" | R |
| (*Sloanea sp. nov.*) "terciopelo" | R |
| *Swietenia macrophylla* G. King "caoba" | R |

Appendix table 5.   List of species in Second-Growth, *cont.*

| Compartments | Relative abundance |
|---|---|
| *Calathea insignis* Peters | O |
| *Corozo oleifera* (H.B.K.) Bailey "corozo" | O |
| *Costus nutans* K. Sch. | O |
| *Renealmia aromatica* (Aubl.) Griseb. | O |
| Understory shrubs | |
| *Clibadium leiocarpum* Steetz "catalina" | D |
| *Acalypha macrostachya* Jacq. | F |
| *Clibadium appressipilum* Blake | F |
| *Clidemia octona* (Bonpl.) L. Wms. | F |
| *Gonzalagunia rudis* Standl. "nigüita" | F |
| *Hamelia axillaris* Sw. | F |
| *Hasseltia floribunda* H.B.K. "parimontón" | F |
| *Isertia haenkeana* DC. | F |
| *Lantana camara* L. "pasorín" | F |
| *Leandra dichotoma* (Don) Cogn. | F |
| *Manihot esculenta* Crantz "yuca" | F |
| *Miconia lacera* (Bonpl.) Naud. | F |
| *Neurolaena lobata* (L.) R. Br. "contragavilán" | F |
| *Piper aduncum* L. | F |
| *Piper darienense* C. DC. "duerme boca" | F |
| *Piper marginatum* Jacq. | F |
| *Rolandra fruticosa* (L.) Kuntze | F |
| *Solanum inerme* Jacq. | F |
| *Solanum rugosum* Dunal | F |
| *Struchium sparganophorum* (L.) Ktze. | F |
| *Tournefortia cuspidata* H.B.K. | F |
| *Triumfetta lappula* L. | F |
| *Vernonia canescens* H.B.K. "hierba de San Juan" | F |
| *Vernonia patens* H.B.K. "lengua de vaca" | F |
| *Abelmoschus moschatus* Medic "naju" | O |
| *Baccharis trinervis* (Lam.) Pers. | O |
| *Bixa orellana* L. "archiote" | O |
| *Cajanus bicolor* DC. "guandú" | O |
| *Callicarpa acuminata* H.B.K. | O |
| *Capsicum baccatum* L. "aji" | O |
| *Capsicum frutescens* L. "aji" | O |
| *Cephaelis tomentosa* (Aubl.) Vahl | O |
| *Clidemia hirta* (L.) D. Don | O |
| *Codiaeum variegatum* (L.) Blume "croton" | O |
| *Corchorus siliquosus* L. | O |
| *Datura candida* (Pers.) Pawq. "campanilla" | O |
| *Indigofera suffruticosa* Mill. | O |
| *Jatropha curcas* L. "coquillo" | O |

Appendix table 5.   List of species in Second-Growth, *cont.*

| Compartments | Relative abundance |
|---|---|
| *Lycianthes synanthera* | O |
| *Malachra alceifolia* Jacq. "malva" | O |
| *Malvastrum americanum* (L.) Torr. | O |
| *Miconia impetiolaris* (Sw.) DC. | O |
| *Miconia rufostellulata* Pitt. | O |
| *Murraya paniculata* (L.) Jacq. "mirto" | O |
| *Piper angustum* Rudge | O |
| *Piper tuberculatum* Jacq. | O |
| *Prockia crucis* L. "huesito" | O |
| *Ricinus communis* L. "higuerillo" | O |
| *Solanum mammosum* L. "tetilla" | O |
| *Solanum quitoensis* Lam. "lulo" | O |
| *Solanum umbellatum* Mill. | O |
| Woody vines | |
| *Byttneria aculeata* Jacq. "espino hueco" | F |
| (*Mucuna bracteata* Dwyer) "ojo de venado" | F |
| *Phryganocydia corymbosa* (Vent.) Bur. | F |
| *Wulffia baccata* (L.f.) Kuntze | F |
| *Allamanda cathartica* L. | O |
| *Calea prunifolia* H.B.K. "escobilla" | O |
| *Calea urticifolia* (Mill.) DC. "escobilla" | O |
| *Gouania lupuloides* (L.) Urb. "jaboncillo" | O |
| *Gouania polygama* (Jacq.) Urb. "jaboncillo" | O |
| *Merremia umbellata* (L.) Hall f. | O |
| *Passiflora auriculata* H.B.K. | O |
| *Passiflora seemannii* Griseb. "guate-quequate" | O |
| *Passiflora suberosa* L. "huevo de gallo" | O |
| *Passiflora vitifolia* H.B.K. "granadillo de monte" | O |
| *Tetrapteryx erythrocarpa* Standl. | O |
| *Vitis tiliaefolia* H. and B. "uva" | O |
| Herbaceous vines | |
| *Cissus rhombifolia* Vahl | F |
| *Cissus sicyoides* L. | F |
| *Dalechampia tiliaefolia* Lam. | F |
| *Desmodium canum* (Gmel.) "pegapega" | F |
| *Eupatorium iresineoides* H.B.K. | F |
| *Eupatorium macrophyllum* L. | F |
| *Eupatorium microstemon* Cass. | F |
| *Eupatorium odoratum* L. | F |
| *Eupatorium sinclairii* Benth. | F |
| *Gurania coccinea* Cogn. "ya te vi" | F |
| *Mikania guaco* H.B.K. | F |
| *Mikania micrantha* H.B.K. | F |

Appendix table 5.   List of species in Second-Growth, *cont.*

| Compartments | Relative abundance |
|---|:---:|
| *Momordica charantia* L. "balsamino" | F |
| *Rhynchosia calycosa* Hemsl. | F |
| *Sabicea villosa* H.B.K. | F |
| *Cayaponia sp.* | O |
| *Centrosema plumieri* (Turp.) Benth. | O |
| *Chaetocalyx latisiliqua* (Desv.) Benth. | O |
| *Chamissoa altissima* (Jacq.) H.B.K. | O |
| *Cissampelos pareira* L. | O |
| *Cissampelos tropaeolifolia* DC. | O |
| *Desmodium barbatum* Benth. | O |
| *Dioscorea polygonoides* Humb. and Bonpl. | O |
| *Gurania seemanniana* Cogn. "ya te vi" | O |
| *Luffa cylindrica* (L.) Roem. "estopa" | O |
| *Manettia coccinea* (Aubl.) Willd. | O |
| *Melanthera aspera* (Jacq.) L. Rich. | O |
| *Passiflora biflora* Lam. | O |
| *P. foetida* L. | O |
| *P. panamensis* Killip | O |
| *Rhynchosia pyramidalis* (Lam.) Urb. "peronilla" | O |
| *Teramnus uncinatus* Sw. | O |
| *Tournefortia angustifolia* R. and P. | O |
| *Vigna vexillata* A. Rich | O |
| *Arbus precatorius* L. "peronilla" | R |
| Canopy trees | |
| *Cecropia peltata* L. "guarumo" | D |
| *Didymopanax morototoni* (Aubl.) Dec. and Pl. | D |
| *Heliocarpus popayanensis* H.B.K. "majagua" | D |
| *Muntingia calabura* L. "perequito" | D |
| *Ochroma pyramidale* (Cav.) Urb. "balsá" | D |
| *Solanum diversifolium* Schlecht. | D |
| *Trema micranta* (L.) Blume "capulin" | D |
| *Carica papaya* L. "papaya" | F |
| *Casearia nitida* (L.) Jacq. "raspa lengua" | F |
| *Cochlospermum vitifolium* (Willd.) Spreng. "poroporo" | F |
| *Cordia alliodora* (R. and P.) R. and S. "laurél" | F |
| *Guazuma ulmifolia* Lam. "gúacimo" | F |
| *Hampea appendiculata* (J.D. Sm.) Standl. | F |
| *Luehea speciosa* Willd. "gúacimo" | F |
| *Piper taboganum* C. DC. | F |
| *Psidium guajava* L. "guayaba" | F |
| *Triplaris cumingiana* Fisch. and Mey. "vara santa" | F |
| *Vismia baccifera* (L.) Tr. and Pl. "sangrillo" | F |

Appendix table 5.    List of species in Second-Growth, *cont.*

| Compartments | Relative abundance |
|---|:---:|
| *V. latifolia* (Aubl.) Choisy "sangrillo" | F |
| *Zanthoxylum panamensis* P. Wils. "arcabú" | F |
| *Z. setulosum* P. Wils. "tachuelo" | F |
| *Annona muricata* L. "guanábana" | O |
| *Byrsonima crassifolia* (L.) H.B.K. "nance" | O |
| *Cecropia longipes* Pitt. "guarumo" | O |
| *Cecropia obtusifolia* Bertol. "guarumo" | O |
| *Clusia spp.* "copey" | O |
| *Cornutia grandiflora* (Schlecht. and Cham.) Schauer "palo cuadrado" | O |
| *Crescentia cujete* L. "merique" | O |
| *Gliricidia sepium* (Jacq.) Steud. "mata ratón" | O |
| *Guettarda odorata* (Jacq.) Lam. "huesito negro" | O |
| *Helicteres guazumaefolia* H.B.K. | O |
| *Margaritaria nobilis* L.f. | O |
| *Miconia argentea* (Sw.) DC. | O |
| *Miconia prasina* (Sw.) DC. | O |
| *Persea americana* Mill. "aguacate" | O |
| *Piper auritum* H.B.K. | O |
| *Spondias purpurea* L. "ciruela" | O |
| *Trichospermum mexicanum* (DC.) Baill. | O |

Appendix table 6.   DBH and names of trees in the Tropical Moist forest (wet season) as shown in figure 2.8.

| | Replication 1 | | | | Replication 2 | | |
|---|---|---|---|---|---|---|---|
| Number | DBH (cm) | Spanish name | Scientific name | Number | DBH (cm) | Spanish name | Scientific name |
| 1 | 18.03 | Carasumar | — | 1 | 21.59 | Guagara | *Sabal alleni* |
| 2 | 12.19 | Tortomito de monte | — | 2 | 20.07 | Guagara | *Sabal alleni* |
| 3 | 23.88 | Carasumar | — | 3 | 40.13 | Guajao | — |
| 4 | 13.46 | Hormiguero | — | 4 | 14.73 | Siete cueros | *Machaerium capote* |
| 5 | 20.07 | Guagara | *Sabal alleni* | 5 | 12.19 | Gorgojo | *Cupania americana* |
| 6 | 21.84 | Guagara | *Sabal alleni* | 6 | 10.16 | Carbonero | — |
| 7 | 18.03 | Guagara | *Sabal alleni* | 7 | 10.67 | Guarumo | *Cecropia sp.* |
| 8 | 21.59 | Guagara | *Sabal alleni* | 8 | 12.70 | Guarumo | *Cecropia sp.* |
| 9 | 19.05 | Guagara | *Sabal alleni* | 9 | 12.45 | Tortomito de monte | — |
| 10 | 22.86 | Guagara | *Sabal alleni* | 10 | 17.53 | Dormilona | *Mimosa pigra* |
| 11 | 10.92 | Carasumar | — | 11 | 19.05 | Guarumo | *Cecropia sp.* |
| 12 | 14.99 | Carasumar | — | 12 | 16.26 | Misperrito | — |
| 13 | 11.43 | Cabo Blanco | — | 13 | 14.48 | Vara Santo | *Triplaris sp.* |
| 14 | 10.41 | Carasumar | — | 14 | 16.51 | Parimonton | — |
| 15 | 23.37 | Guayacan | *Tabebuia guayacan* | 15 | 12.45 | Parimonton | — |
| 16 | 20.07 | Guagara | *Sabal alleni* | 16 | 13.46 | Parimonton | — |
| 17 | 59.18 | Caucho | *Castilla elastica* | 17 | 18.80 | Parimonton | — |
| 18 | 16.26 | Guagara | *Sabal alleni* | 18 | 22.35 | Parimonton | — |
| 19 | 114.55 | Roble | *Tabebuia pentaphylla* | 19 | 22.10 | Parimonton | — |
| 20 | 11.18 | Tortomito de monte | — | 20 | 20.57 | Guagara | *Sabal alleni* |

| No. | | | | No. | | | |
|---|---|---|---|---|---|---|---|
| 21 | 15.49 | Guarumo | *Cecropia sp.* | 21 | 19.05 | Guagara | *Sabal alleni* |
| 22 | 10.41 | Membrillo | *Gustavia superba* | 22 | 20.07 | Guagara | *Sabal alleni* |
| 23 | 25.66 | Berbasillo | — | 23 | 20.07 | Guagara | *Sabal alleni* |
| 24 | 48.28 | Caucho | *Castilla elastica* | 24 | 10.92 | Cabo Blanco | — |
| 25 | 17.78 | Tortomito de monte | — | 25 | 22.35 | Carasumar | — |
| 26 | 29.97 | Cedro macho | — | 26 | 115.32 | Cuipo | *Cavanillesia plantanifolia* |
| 27 | 10.67 | Guamoguere | — | 27 | 20.83 | Guarumo | *Cecropia sp.* |
| 28 | 17.37 | Guagara | *Sabal alleni* | 28 | 16.51 | Guabito | — |
| 29 | 16.51 | Guagara | *Sabal alleni* | 29 | 11.18 | Cabimo macho | — |
| 30 | 11.68 | Churimito | — | 30 | 21.84 | Guagara | *Sabal alleni* |
| 31 | 10.41 | Charasumar | — | 31 | 24.64 | Carasumar | — |
| 32 | 36.07 | Carasumar | — | 32 | 10.92 | Membrillo | *Gustavia superba* |
| 33 | 13.72 | Misperrito | — | 33 | 17.27 | Membrillo | *Gustavia superba* |
| 34 | 22.86 | Guagara | *Sabal alleni* | 34 | 16.00 | Cabimo | — |
| 35 | 13.21 | Berbasillo | — | 35 | 11.18 | Cabo Macho | — |
| 36 | 17.53 | Membrillo | *Gustavia superba* | 36 | 19.81 | Guagara | *Sabal alleni* |
| 37 | 19.30 | Tortomito de monte | — | 37 | 44.45 | Caucho | *Castilla elastica* |
| 38 | 21.59 | Tapalisa | *Chomelia chambersii* | 38 | 14.99 | Berbasillo | — |
| 39 | 16.26 | Tortomito de monte | — | 39 | 50.04 | Gorgojo | *Cupania americana* |
| 40 | 16.26 | Tortomito de monte | — | 40 | 23.11 | Guagara | *Sabal alleni* |
| 41 | 21.35 | Guagara | *Sabal alleni* | 41 | 23.61 | Gorgojo | *Cupania americana* |
| 42 | 19.56 | Guagara | *Sabal alleni* | 42 | 20.57 | Gorgojo | *Cupania americana* |
| 43 | 17.02 | Guagara | *Sabal alleni* | 43 | 17.78 | Guagara | *Sabal alleni* |
| 44 | 12.95 | Tortomito de monte | — | 44 | 67.56 | Tuqueso | *Licania sp.?* |
| 45 | 23.11 | Guagara | *Sabal alleni* | 45 | 18.03 | Chimito | — |
| 46 | 21.59 | Guagara | *Sabal alleni* | 46 | 22.86 | Naranjita | *Swartzia simplex* |
| 47 | 19.56 | Guagara | *Sabal alleni* | 47 | 18.03 | Fruta mono macho | — |

Appendix table 6.   DBH and names of trees in the Tropical Moist forest (wet season) as shown in figure 2.8, *co*

| | Replication 1 | | | | Replication 2 | | |
|---|---|---|---|---|---|---|---|
| Number | DBH (cm) | Spanish name | Scientific name | Number | DBH (cm) | Spanish name | Scientific name |
| 48 | 19.05 | Guagara | *Sabal alleni* | 48 | 19.81 | Guayacan | *Tabebuia guayacan* |
| 49 | 20.83 | Guagara | *Sabal alleni* | 49 | 25.40 | Membrillo | *Gustavia superba* |
| 50 | 19.05 | Tortomito de monte | — | 50 | 12.70 | Tortomito de monte | |
| 51 | 12.70 | Tortomito de monte | — | 51 | 11.43 | Alfajia | *Trichillia sp.* |
| 52 | 27.94 | Naranjita | *Swartzia simplex* | 52 | 18.22 | Cabo Blanco | |
| 53 | 17.78 | Guagara | *Sabal alleni* | 53 | 16.76 | Cabo Blanco | |
| 54 | 39.12 | Coco bolo | — | 54 | 11.94 | Tortomito de monte | |
| 55 | 17.78 | Guagara | *Sabal alleni* | 55 | 10.41 | Misperrito | |
| 56 | 18.54 | Guagara | *Sabal alleni* | 56 | 12.19 | Membrillo | *Gustavia superba* |
| 57 | 16.26 | Cabo Blanco | — | 57 | 35.56 | Cedro Macho | |
| 58 | 18.29 | Guagara | *Sabal alleni* | 58 | 222.25 | Cuipo | *Cavanillesia plantanifolia* |
| 59 | 22.86 | Guagara | *Sabal alleni* | 59 | 13.72 | Yaya Montanero | |
| 60 | 12.45 | Coratu | — | 60 | 10.67 | Cabo Blanco | — |
| 61 | 11.43 | Naranjillo Macho | | 61 | 12.45 | Yaya Laguna | — |
| | | | | 62 | 17.27 | Membrillo | *Gustavia superba* |
| | | | | 63 | 20.32 | Membrillo | *Gustavia superba* |
| | | | | 64 | 20.57 | Membrillo | *Gustavia superba* |

Appendix table 7.   DBH of trees in the Premontane Wet forest
as shown in figure 4.7.

| Replication 1 | | Replication 2 | |
| --- | --- | --- | --- |
| Number | DBH (cm) | Number | DBH (cm) |
| 1 | 69.34 | 1 | 10.92 |
| 2 | 28.45 | 2 | 10.16 |
| 3 | 30.73 | 3 | 83.65 |
| 4 | 17.78 | 4 | 11.18 |
| 5 | 21.84 | 5 | 41.91 |
| 6 | 15.75 | 6 | 14.99 |
| 7 | 14.22 | 7 | 41.66 |
| 8 | 33.27 | 8 | 17.78 |
| 9 | 15.48 | 9 | 14.99 |
| 10 | 17.53 | 10 | 40.64 |
| 11 | 26.67 | 11 | 10.92 |
| 12 | 13.97 | 12 | 10.16 |
| 13 | 15.75 | 13 | 66.55 |
| 14 | 52.32 | 14 | 22.61 |
| 15 | 13.97 | 15 | 12.70 |
| 16 | 20.32 | 16 | 22.86 |
| 17 | 12.70 | 17 | 14.73 |
| 18 | 15.24 | 18 | 12.45 |
| 19 | 16.76 | 19 | 18.03 |
| 20 | 21.59 | 20 | 18.03 |
| 21 | 34.54 | 21 | 28.70 |
| 22 | 40.89 | 22 | 17.27 |
| 23 | 13.97 | 23 | 7.37 |
| 24 | 12.19 | 24 | 21.84 |
| 25 | 11.18 | 25 | 13.72 |
| 26 | 18.54 | 26 | 12.19 |
| 27 | 30.23 | 27 | 34.54 |
| 28 | 13.21 | 28 | 11.18 |
| 29 | 13.97 | 29 | 24.13 |
| 30 | 95.25 | 30 | 28.96 |
| 31 | 11.18 | 31 | 10.16 |
| 32 | 12.45 | 32 | 10.41 |
| 33 | 48.51 | 33 | 21.59 |
| 34 | 22.35 | 34 | 11.68 |
| 35 | 11.43 | 35 | 12.45 |
| 36 | 14.22 | 36 | 13.72 |
| 37 | 23.62 | 37 | 10.41 |
| 38 | 28.70 | 38 | 10.67 |
| 39 | 30.73 | 39 | 14.99 |
| 40 | 21.84 | 40 | 12.45 |

Appendix table 7.   DBH of trees in the Premontane Wet forest as shown in figure 4.7, *cont.*

| Replication 1 | | Replication 2 | |
|---|---|---|---|
| Number | DBH (cm) | Number | DBH (cm) |
| 41 | 23.62 | 41 | 17.02 |
| 42 | 10.67 | 42 | 13.97 |
| 43 | 22.10 | 43 | 19.56 |
| 44 | 52.07 | 44 | 16.00 |
| 45 | 21.34 | 45 | 54.61 |
| 46 | 10.92 | 46 | 16.76 |
| 47 | 47.24 | 47 | 17.02 |
| 48 | 10.41 | 48 | 29.99 |
| 49 | 23.62 | 49 | 14.99 |
| 50 | 21.84 | 50 | 15.49 |
| 51 | 16.51 | 51 | 21.34 |
| 52 | 12.45 | 52 | 16.26 |
| 53 | 59.44 | 53 | 14.22 |
| 54 | 37.34 | 54 | 15.24 |
| 55 | 16.76 | 55 | 21.59 |
| 56 | 13.72 | 56 | 21.34 |
| 57 | 28.19 | 57 | 10.92 |
| 58 | 11.43 | 58 | 10.41 |
| 59 | 17.78 | 59 | 19.05 |
| 60 | 10.16 | 60 | 10.41 |
| 61 | 65.53 | 61 | 22.35 |
| 62 | 22.86 | 62 | 11.93 |
| 63 | 20.57 | 63 | 10.92 |
| 64 | 11.18 | 64 | 10.41 |
| 65 | 14.22 | 65 | 41.40 |
| 66 | 12.95 | 66 | 42.67 |
| 67 | 23.88 | 67 | 11.94 |
| 68 | 10.16 | 68 | 25.40 |
| | | 69 | 44.45 |
| | | 70 | 10.41 |
| | | 71 | 11.94 |
| | | 72 | 14.99 |
| | | 73 | 15.49 |
| | | 74 | 15.75 |
| | | 75 | 18.03 |
| | | 76 | 16.26 |
| | | 77 | 13.72 |
| | | 78 | 21.08 |
| | | 79 | 11.68 |
| | | 80 | 14.48 |

Appendix table 8.   DBH of trees in the Mangrove forest as shown in figure 4.7. All trees are Red Mangrove, *Rhizophora brevistyla*.

| Replication 1 | | Replication 2 | | Replication 2, *cont*. | |
|---|---|---|---|---|---|
| Number | DBH (cm) | Number | DBH (cm) | Number | DBH (cm) |
| 1 | 13 | 1 | 13 | 43 | 15 |
| 2 | 10 | 2 | 13 | 44 | 35 |
| 3 | 13 | 3 | 15 | 45 | 19 |
| 4 | 13 | 4 | 14 | 46 | 10 |
| 5 | 27 | 5 | 13 | 47 | 46 |
| 6 | 11 | 6 | 10 | | |
| 7 | 10 | 7 | 12 | | |
| 8 | 19 | 8 | 10 | | |
| 9 | 15 | 9 | 10 | | |
| 10 | 12 | 10 | 19 | | |
| 11 | 10 | 11 | 14 | | |
| 12 | 14 | 12 | 12 | | |
| 13 | 26 | 13 | 12 | | |
| 14 | 11 | 14 | 15 | | |
| 15 | 15 | 15 | 17 | | |
| 16 | 15 | 16 | 12 | | |
| 17 | 13 | 17 | 19 | | |
| 18 | 35 | 18 | 22 | | |
| 19 | 12 | 19 | 13 | | |
| 20 | 25 | 20 | 14 | | |
| 21 | 12 | 21 | 16 | | |
| 22 | 19 | 22 | 14 | | |
| 23 | 10 | 23 | 21 | | |
| 24 | 22 | 24 | 13 | | |
| 25 | 30 | 25 | 34 | | |
| 26 | 11 | 26 | 13 | | |
| 27 | 19 | 27 | 15 | | |
| 28 | 23 | 28 | 22 | | |
| 29 | 31 | 29 | 20 | | |
| 30 | 18 | 30 | 15 | | |
| 31 | 11 | 31 | 34 | | |
| 32 | 33 | 32 | 38 | | |
| 33 | 19 | 33 | 29 | | |
| 34 | 24 | 34 | 21 | | |
| 35 | 24 | 35 | 49 | | |
| 36 | 13 | 36 | 33 | | |
| 37 | 49 | 37 | 11 | | |
| 38 | 63 | 38 | 12 | | |
| 39 | 22 | 39 | 14 | | |
| 40 | 29 | 40 | 26 | | |
| 41 | 38 | 41 | 16 | | |
| 42 | 12 | 42 | 10 | | |

## Appendix table 9. DBH and names of trees in the Riverine forest as shown in figure 4.7.

| | Replication 1 | | | | Replication 2 | | |
|---|---|---|---|---|---|---|---|
| Number | DBH (cm) | Spanish name | Scientific name | Number | DBH (cm) | Spanish name | Scientific name |
| 1 | 74.17 | Cativo | *Prioria copafeira* | 1 | 62.74 | Cativo | *Prioria copafeira* |
| 2 | 119.38 | Cativo | *Prioria copafeira* | 2 | 24.13 | Cativo | *Prioria copafeira* |
| 3 | 23.88 | Alfajia | *Trichilia sp.* | 3 | 49.02 | Parimonton | ? |
| 4 | 31.50 | Perico | ? | 4 | 18.80 | (dead tree) | ? |
| 5 | 192.53 | Espavé | *Anacardium excelsum* | 5 | 16.76 | Cativo | *Prioria copafeira* |
| 6 | 14.99 | Dead tree | ? | 6 | 11.43 | Carbonero | ? |
| 7 | 88.90 | Cativo | *Prioria copafeira* | 7 | 24.64 | Guayacan | *Tabebuia guayacan* |
| 8 | 10.67 | Carnobero | ? | 8 | 68.68 | Cativo | *Prioria copafeira* |
| 9 | 10.67 | Yaya negro | ? | 9 | 12.19 | Cativo | *Prioria copafeira* |
| 10 | 11.43 | Punula | *Quararibaea cf. asterolepis* | 10 | 30.48 | Cativo | *Prioria copafeira* |
| 11 | 10.67 | Punula | *Quararibaea cf. asterolepis* | 11 | 40.64 | Pialoa | ? |
| 12 | 80.26 | Cativo | *Prioria copafeira* | 12 | 14.22 | Alfajia | *Trichilia sp.* |
| 13 | 54.36 | Cativo | *Prioria copafeira* | 13 | 11.43 | Yaya laguna | (Family—Anonaceae) |
| 14 | 47.24 | Cativo | *Prioria copafeira* | 14 | 14.48 | Cativo | *Prioria copafeira* |
| 15 | 11.18 | Cauchillo | *Castilla sp.* | 15 | 15.49 | Alfajia | *Trichilia sp.* |
| 16 | 37.08 | Cativo | *Prioria copafeira* | 16 | 10.16 | Cativo | *Prioria copafeira* |
| 17 | 42.67 | Cativo | *Prioria copafeira* | 17 | 10.92 | Tugueso | ? |
| 18 | 15.75 | Alfajia | *Trichilia sp.* | 18 | 15.24 | Alfajia | *Trichilia sp.* |
| 19 | 23.88 | Cativo | *Prioria copafeira* | 19 | 44.96 | Cativo | *Prioria copafeira* |

| No. | Value | Common name | Scientific name |
|---|---|---|---|
| 20 | 29.46 | Cativo | *Prioria copafeira* |
| 21 | 35.05 | Cativo | *Prioria copafeira* |
| 22 | 17.02 | Chunga | *Astrocarum standleyanum* |
| 23 | 37.34 | Cutaro | *Swartzia sp.* |
| 24 | 33.53 | Cativo | *Prioria copafeira* |
| 25 | 22.35 | Cativo | *Prioria copafeira* |
| 26 | 16.51 | Cativo | *Prioria copafeira* |
| 27 | 11.18 | Cauchillo | *Castilla sp.* |
| 28 | 17.53 | Cativo | *Prioria copafeira* |
| 29 | 15.85 | Chira | *Platymyscium sp.* |

| No. | Value | Common name | Scientific name |
|---|---|---|---|
| 20 | 14.73 | Yaya laguna | (Family—Anonaceae) |
| 21 | 97.28 | Cativo | *Prioria copafeira* |
| 22 | 14.48 | Cutaro | *Swartzia sp.* |
| 23 | 23.11 | Cuadero | *Vitex sp.* |
| 24 | 10.92 | Fruta de pava | ? |
| 25 | 16.76 | Cutaro | *Swartzia sp.* |
| 26 | 11.18 | Sega | ? |
| 27 | 13.21 | Carbonero | ? |
| 28 | 16.00 | Cativo | *Prioria copafeira* |
| 29 | 14.22 | Cutaro | *Swartzia sp.* |
| 30 | 22.86 | Alfajia | *Trichilia sp.* |
| 31 | 111.00 | Cativo | *Prioria copafeira* |
| 32 | 119.63 | Cativo | *Prioria copafeira* |
| 33 | 10.92 | Membrillo | *Gustavia superba* |
| 34 | 80.52 | Cativo | *Prioria copafeira* |
| 35 | 12.95 | Cativo | *Prioria copafeira* |
| 36 | 10.67 | Tugueso | *Licania sp.* ? |
| 37 | 23.88 | Cativo | *Prioria copafeira* |
| 38 | 33.53 | Cativo | *Prioria copafeira* |
| 39 | 78.74 | Cativo | *Prioria copafeira* |
| 40 | 140.97 | Suela | *Pterocarpus sp.* |

# LIST OF
# WORKS CITED

Allee, W. C. 1926. Distribution of animals in a tropical rain-forest with relation to environmental factors. *Ecology* 7:445-468.

Attiwell, P. M. 1966. The chemical composition of rainwater in relation to cycling of nutrients in mature Eucalyptus forest. *Plant and Soil* 24:390-406.

Bartholomew, W. V., J. Meyer, and H. Laudelout. 1953. *Mineral Nutrient immobilization under forest and grass fallow in the Yangambi (Belgian Congo) region.* Publ. Inst. Nat. Etude Agron. Congo Belge, Ser. Sci. no. 57.

Bartlett, H. H. 1955, 1957, and 1961. Fire in relation to primitive agriculture and grazing in the tropics. Annotated bibliography, vol. 1, vol. 2, vol. 3. Ann Arbor, Dept. of Botany, Univ. of Michigan.

Beyers. R., M. H. Smith, J. B. Gentry, and L. L. Ramsey. 1970. Constancy of atomic ratios of six elements in three species of small mammals. *Acta Theriologica* 16:203-211.

Birot, P. 1968. *The cycle of erosion in different climates.* Berkley: University of California Press.

Blue, W. G., C. B. Ammerman, J. M. Loaiza, and J. F. Gamble. 1969. Compositional analysis of soils, forages, and cattle tissues from beef producing areas of eastern Panama. In *Symposium on sea-level canal bioenvironmental studies.* IOCS memorandum BMI-24, report 11.

Boggess, W. R. 1956. Amount of throughfall and stemflow in shortleaf pine plantations as related to rainfall in the open. *Illinois Acad. Sci. Trans.* 48:55-61.

Bormann, F. H., G. E. Likens, and J. S. Eaton. 1969. Biotic regulation of particulate and solution losses from a forest ecosystem. *BioScience* 19:600-610.

Bowen, H. J. M. 1966. *Trace elements in biochemistry.* London: Academic Press.

Bray, J. R. 1961. Measurement of leaf utilization as an index of minimum level of primary consumption. *Oikos* 12:70-74.

Brinkmann, W. L. F., and A. dos Santos. 1971. Natural waters in Amazonia. v. Soluble magnesium properties. *Turrialba* 21:459-465.

Budowski, G. 1966. La influencia del hombre precolombino en la vegetacion Tropical Americana. *XXXVI Congress Int. d. Americanistas* 1:115-118.

Burns, L. A. 1970. Analog simulation of rain forest, with high-low pass filters and a programmatic spring pulse. In *A tropical rain forest,* ed. H. T. Odum and R. F. Pigeon, pp. 1284-1289. USAEC.

Cain S. A., and G. M. d'O. Castro. 1959. *Manual of vegetation analysis.* New York: Harper and Bros.

Cain, S. A., G. M. d'O Castro, J. M. Pires, and N. T. da Silva. 1956. Application of some phytosociological techniques to Brazilian rain forest. *Amer. J. Botany* 43:911-941.

Clegg, A. G. 1963. Rainfall interception in a tropical forest. *Caribbean Forester* 24:75-79.

Cole, D. W., S. P. Gessel, and S. F. Dice. 1968. Distribution and cycling of nitrogen, phosphorus, potassium, and calcium in a second-growth Douglas-fir ecosystem. In *Symposium of primary productivity and mineral cycling in natural ecosystems,* pp. 197-232. Orano: Univ. of Maine Press.

Comrey, A. L. 1973. *A first course in factor analysis.* New York: Academic Press.

Conklin, H. C. 1963. *The study of shifting cultivation.* Panamerican Union Studies and Monographs, no. 6.

Cooper, C. F. 1969. Ecosystem models in watershed management. In *The ecosystem concept in national resource management,* ed. G. M. Van Dyne, pp. 309-24. New York: Academic Press.

Coulter, J.K. 1950. Organic matter in Malayan soils. *Malayan Forester* 13:189-202.

Crossley, D. A., Jr., and H. F. Howden. 1961. Insect-vegetation relationships in an area contaminated by radioactive wastes. *Ecology* 42:302-317.

Cuatrecasas, J. 1958. Introduction al estudio de los manglares. *Bol. Soc. Bot. d. Mexico* 23:84-98.

Cunningham, R. K. 1963. The effect of clearing a tropical forest soil. *J. Soil Sci.* 14:334-344.

Davis, D. E., and F. B. Golley. 1963. *Principles in mammalogy.* New York: Reinhold Publ. Corp.

de la Cruz, A. 1964. A preliminary study of organic detritus in a tropical forest ecosystem. *Rev. Biol. Trop.* 12:175-185.

Deevey, E. S., Jr. 1970. Mineral cycles. *Sci. Amer.* 223:149-158.

Dils, R. E. 1957. *A guide to the Coweeta Hydrologic Laboratory.* Southeastern Forest Exp. Station, Asheville, N. C. Misc. Publ.

Dommerques, Y. 1963. Les cycles biogéochimiques des éléments minéraux dans les formations tropicales. *Bois et foréts des tropiques* 87: 10-25.

Drees, E. M. 1954. The minimum area in tropical rain forest with special reference to some types in Bangka (Indonesia). *Vegetatio* 5/6:517-523.

Dudal, R. 1963. Dark clay soils of tropical and subtropical regions. *Soil Science* 95:264-270.

Duever, Andrea J. 1967. Trophic dynamics of reptiles in terms of the community food web and energy intake. M.S. thesis, University of Georgia.

Edwards, E. E., and W. D. Rasmussen. 1942. *A bibliography on the agriculture of the American Indians.* USDA Misc. Publ. no. 447.

Edwards, C. A. 1967. Relationships between weights, volumes, and numbers of soil animals. In *Progress in Soil Biology,* ed. O. Graff and J. E. Satchell, Amsterdam. pp. 1-10.

Edwards, H. W. 1971. Impact on man of environmental contamination caused by lead. NSF Grant Interim Report, Colorado State Univ.

Eisenmann, E. 1955. The species of middle American birds. *Trans. Linnean Soc. N. Y.* 7:1-128.

Emanuelsson, A., E. Eriksson, and H. Egner. 1954. Composition of atmospheric precipitation in Sweden. *Tellus* 6:261-267.

Englemann, M. D. 1961. The role of soil arthropods in the energetics of an old-field community. *Ecol. Monog.* 31:221-238.

Ewel, J. 1971. Biomass changes in early tropical succession. *Turrialba* 21:110-112.

Freise, F. 1936. Das Binnenklima von Urwäldern im subtropischen Brasilien. *Petermans Mitt.* 82:301-307.

Gamble, J. F., R. Ah Chu, and J. G. A. Fiskell. 1969. Soils and agriculture of eastern Panama and northwestern Colombia. In *Symposium on sea-level canal bioenvironmental studies.* IOCS memorandum BMI-24, report 8.

Giglioli, M. E. C., and I. Thornton. 1965. The mangrove swamps of Keneba, lower Gambia River basin. I. Descriptive notes on the climate, the mangrove swamps and the physical composition of their soils. *J. Appl. Ecol.* 2:81-103.

Gilbert, F. A. 1957. *Mineral nutrition and the balance of life.* Norman: Univ. Oklahoma Press.

Golley, F. B. 1965. Structure and function of an old-field broomsedge community. *Ecol. Monog.* 35:113-131.

Golley, F. B. 1969. Caloric value of wet tropical forest vegetation. *Ecology* 50:517-519.

Golley, F. B. 1971. Impact of small mammals on primary production. AIBS meeting, Ft. Collins, Colorado. Mimeographed.

Golley, F. B. 1972a. Summary. In *Tropical Ecology,* with an emphasis on organic production. P. M. Golley and F. B. Golley, compilers, pp. 407-413. Athens, Georgia: International Society of Tropical Ecology.

Golley, F. B. 1972b. Energy flux in ecosystems. In *Ecosystem structure and function,* ed. J. A. Wiens, pp. 69-90. Corvallis: Oregon State Univ. Press.

Golley, F. B., H. T. Odum, and R. F. Wilson. 1962. The structure and metabolism of a Puerto Rican red mangrove forest in May. *Ecology* 43:9-19.

Gorham, E. 1961. Factors influencing supply of major ions to inland waters, with special reference to the atmosphere. *Geol. Soc. Amer. Bull.* 72:795-840.

Gray, H. E., and A. E. Treloar. 1933. On the enumeration of insect populations by the method of net collection. *Ecology* 14:356-367.

Greenland, D. J., and J. M. L. Kowal. 1960. Nutrient content of the moist tropical forest of Ghana. *Plant and Soil* 12(2):154-174.

Greig-Smith, P. 1964. *Quantitative plant ecology.* 2nd ed. London: Butterworths.

Grubb, P. J., J. R. Lloyd, T. D. Pennington, and T. C. Whitmore. 1963. A comparison of montane and lowland rain forest in Ecuador. I. The forest structure, physiognomy and floristics. *J. Ecol.* 51:567-601.

Guzman, L. E. 1956. Farming and farmlands in Panama. Dept. Geography res. paper, 44, Univ. of Chicago.

Harrison, J. L. 1962. The distribution of feeding habits among animals in a tropical rain forest. *J. Anim. Ecol.* 31:53-64.

Harrison, J. L. 1969. The abundance and population density of mammals in Malayan lowland forests. *Malay. Nat. J.* 22:174-178.

Heatwole, H., and O. J. Sexton. 1966. Herpetofaunal comparisons between two climatic zones in Panama. *Amer.Midl. Nat.* 75(1):45-60.

Helvey, J. D. and J. H. Patric. 1965. Canopy and litter interception of rainfall by hardwoods of eastern United States. *Water Resources Res.* 1(2):193-206.

Holdridge, L. R. 1964. Ecology. In TreCom Techn. rept. 63-72. Environmental Survey, Darien Province, Rep. Panama. 1962. U. S. Army Transportation Res. Com., Ft. Eustice, Va. pp. 165-333.

Holdridge, L. R. 1971. *Forest environments in tropical life zones; a pilot study.* New York: Pergamon Press.

Holdridge, L. R., and G. Budowski. 1956. Report of an ecological survey of the republic of Panama. *Caribbean Forester* 17:92-110.

Hopkins, B. 1966. Vegetation of the Olokemeji forest reserve, Nigeria. IV. The litter and soil with special reference to their seasonal changes. *J. Ecology* 54:687-703.

IOCS-FD-31. 1967. Interim geologic report, number II, route 17. Engineering feasibility studies, Atlantic-Pacific Interoceanic Canal. Office Interoceanic Canal Studies.

IOCS-FD-59. 1968. Hydrology Route 17. Final report. Engineering Feasibility Studies, Atlantic-Pacific Interoceanic Canal. Office Interoceanic Canal Studies, Panama.

IOCS-FD-60. 1968. Subsurface geology data collection, Raw data— laboratory results. Part I. Physical properties test results, route 17. Engineering Feasibility Studies, Atlantic-Pacific Interoceanic Canal. Office Interoceanic Canal Studies.

Iwatsubo, G., and T. Tsutsumi. 1968. On the amount of plant nutrients supplied to the ground by rainwater in adjacent open plot and forests. (III). On the amount of plant nutrients contained in run-off water. *Bull. Kyoto Univ. For.* 40:140-156.

Janzen, D. H. 1967. Synchronization of sexual reproduction of trees within the dry season in central America. *Evolution* 21(3):620-637.

Janzen, D. H., and T. W. Schoener. 1968. Differences in insect abundance and diversity between wetter and drier sites during a tropical dry season. *Ecology* 49(1):96-110.

Jenny, H., S. P. Gessel, and F. T. Bingham. 1949. Comparative study of decomposition rates of organic matter in temperate and tropical regions. *Soil Sci.* 68:419-432.

Johnson, D. R. 1966. Diet and estimated energy assimilation of three Colorado lizards. *Amer. Midl. Nat.* 76(2):504-509.

Jordan, C. F. 1971. Productivity of a tropical forest and its relation to a world pattern of energy storage. *J. Ecol.* 59:127-142.

Jordan, C. F., and J. R. Kline. 1972. Mineral cycling: some basic concepts and their application in a tropical rain forest. *Ann. Rev. Ecology and Systematics* 3:33-50.

Jordan, C. F., J. R. Kline, and D. S. Sasscer. 1972. Relative stability of mineral cycles in forest ecosystems. *Amer. Nat.* 106:237-253.

Jung, G. 1969. Cycles biogéochimiques dans un écosystème de région tropicale sèche *Acacia albida* (Del.) Sol ferrugineux tropical peu lessivé (Dior.). *Oecol. Plant.* 4:195-210.

Junge, C. E., and R. T. Werby. 1958. The concentration of chloride, sodium, potassium, calcium, and sulfate in rain water over the United States. *J. Meteorology* 15(5):417-425.

Karr, J. R. 1971. Structure of avian communities in selected Panama and Illinois habitats. *Ecol. Monog.* 41:207-233.

Kauffman, R. G., L, E. St. Clair, and R. J. Reber. 1963. *Ovine Myology.* Univ. of Ill. Agr. Experiment Station Bull. 698:1-54.

Kehoe, R. A. 1961. The normal metabolism of lead. *J. Ray Inst. Publ. Health and Hyg.* 24:81-96.

Kellman, M. C. 1970. *Secondary plant succession in tropical montane Mindanao.* Canberra: Australian Natl. Univ. Press.

Kira, T., H. Ogawa, K. Yoda, and K. Ogino. 1964. Primary production by a tropical rain forest of southern Thailand. *Bot. Mag., Tokyo.* 77:428-429.

Kira, T., H. Ogawa, K. Yoda, and K. Ogino. 1967. Comparative ecological studies on three main types of forest vegetation in Thailand. IV. Dry matter production, with special reference to the Khao Chong rain forest. *Nature and Life in S. E. Asia* 5:149-174.

Klinge, H., and W. A. Rodrigues. 1968a. Litter production in an area of Amazonian Terra Firme forest. Part I. Litter-fall, organic carbon and total nitrogen contents of litter. *Amazonia* 1:287-302.

Klinge, H., and W. A. Rodrigues. 1968b. Litter production in an area of Amazonian Terre Firme forest. Part II. Mineral nutrient content of the litter. *Amazonia* 1:303-310.

Krebs, J. E. 1972. Comparison of soils under agriculture and forests in San Carlos, Costa Rica. M. S. thesis, University of Georgia.

Lamb, F. B. 1953. The forests of Darien, Panama. *Caribbean Forester* 14:128-135.

Lamberti, A. 1969. *Contribuicão ao conhecimento da ecologia das plantas do Manguezal de Itanhaém.* Univ. São Paulo Bol. no. 317, Bot. no. 23.

Laudelout, H., and J. Meyer. 1954. Les cycles d'élèments minéraux et de matière organique en forêt équatoriale Congolaise. *Trans. Fifth Int. Cong. Soil Sci.* 2:267-272.

Lawson, E. R. 1967. Throughfall and stemflow in a pine-hardwood stand in the Ouachita Mountains of Arkansas. *Water Resources Res.* 3:731-735.

Leopold, L. B., M. G. Wolman, and J. P. Miller. 1964. *Fluvial processes in geomorphology.* San Francisco: W. H. Freeman.

Likens, G. E., F. H. Bormann, N. M. Johnson, and R. S. Pierce. 1967. The calcium, magnesium, potassium, and sodium budgets for a small forested ecosystem. *Ecology* 48(5):772-785.

Lomnicki, A., A. Kosior, and T. Kazmierczak. 1965. Ocena suchej masy uszkodzén, dokonanych przez roślinozerćow w runie lasu bukowego *(Fagetum carpaticum) Ekologia Polska B.* 11:61-67.

Lyon, T. L., and H. O. Buckman. 1949. *The nature and properties of soils.* 4th ed. New York: Macmillan Co.

MacLeish, K., H. H. Hennefrund, and M. G. Lacy. 1940. *Anthropology and agriculture: selected references on agriculture in primitive cultures.* USDA Bureau Agric. Econ., Bibliography 89.

Madge, D. S. 1965. Leaf fall and litter disappearance in a tropical forest. *Pedobiologica* 5:273-288.

Martini, J. A., R. Ah Chu, P. N. Lezcano, and J. W. Brown. 1960. Forest soils of Darien Province, Panama. *Tropical Woods* 112:28-39.

Matthew, G. E. 1950. *Rainfall and Runoff of the Gatun Lake Watershed 1907-1948.* The Panama Canal Engin. Const. Bureau, Meterological and Hydrographic Branch.

Mayo Melendez, E. 1965. Algunas características ecológicas de los bosques inundables de Darién, Panamá, con miras a su posible utilización. *Turrialba* 15:336-347.

McNab, B. K. 1963. A model of the energy budget of a wild mouse. *Ecology* 44(3):521-532.

Misra, R. 1968. Energy transfer along terrestrial food chain. *Tropical Ecology* 9(2):105-118.

Misra, R. 1972. A comparative study of net primary productivity of dry deciduous forest and grassland of Yaranasi, India. In Tropical ecology, with an emphasis on organic production. P. M. Golley and F. B. Golley, compilers, pp. 279-293. Athens, Georgia.

Mohr, E. C. J., and F. A. Van Baren. 1954. *Tropical soils.* New York: Interscience Publ.

Müller, D., and J. Nielsen. 1965. Production brute, pertes par respiration et production nette dans la forêt ombrophile tropicale. *Det Forstlige Forsøgsvaesen i Danmark* 29:69-160.

Nice, M. M. 1938. The biological significance of bird weights. *Birdbanding* 9:1-11.

Nye, P. II. 1961. Organic matter and nutrient cycles under moist tropical forest. *Plant and Soil* 13(4):333-346.

Nye, P. H., and D. J. Greenland. 1960. *The soil under shifting cultivation.* Tech. Comm. no. 51, Commonwealth Bureau Soil Sci.

Nye, P. H., and D. J. Greenland. 1964. Changes in the soil after clearing tropical forest. *Plant and Soil* 21(1):101-112.

Odum, H. T. 1967. *Hydrogen budget and compartments in the rain forest at El Verde, Puerto Rico, pertinent to consideration of tritium metabolism.* IOCS memorandum BMI-2.

Odum, H. T. 1970a. Summary: an emerging view of the ecological system at El Verde. In *A tropical rain forest,* ed. H. T. Odum and R. F. Pigeon, pp. I-191-289. USAEC.

Odum, H. T. 1970b. Rain forest structure and mineral cycling homeostasis. In *A tropical rain forest,* ed. H. T. Odum and R. F. Pigeon, pp. H3-52. USAEC.

Odum, H. T., and R. F. Pigeon. 1970. *A tropical rain forest; a study of irradiation and ecology at El Verde, Puerto Rico.* USAEC, Div. Tech. Inf.

Odum, H. T., W. Abbott, R. K. Sealander, F. B. Golley, and R. F. Wilson. 1970. Estimates of chlorophyll and biomass of the Tabonuco forest of Puerto Rico. In *A tropical rain forest,* ed. H. T. Odum and R. F. Pigeon, pp. I3-19. USAEC.

Ogawa, H., K. Yoda, and T. Kira. 1961. A preliminary survey of the vegetation of Thailand. *Nature and Life in Southeast Asia* 1:21-157.

Ogawa, H., K. Yoda, K. Ogino, and T. Kira. 1965. Comparative ecological studies on three main types of forest vegetation in Thailand. II. Plant biomass. *Nature and Life in Southeast Asia* 4:49-80.

Olson, J. S. 1963. Energy storage and the balance of producers and decomposers in ecological systems. *Ecology* 44:322-331.

Olson, J. S. 1970. Geographic index of world ecosystems. In *Analysis of temperate forest ecosystems,* ed. D. E. Reichle, pp. 297-304. New York, Heidelberg, Berlin: Springer-Verlag.

Ovington, J. D. 1962. Quantitative ecology and the woodland ecosystem concept. In *Advances in ecological research,* vol. 1, ed. J. B. Cragg, pp. 103-192. London: Academic Press.

Ovington, J. D. 1959. The circulation of minerals in plantations of *Pinus sylvestris* L. *Ann. Bot.* 23:230-239.

Ovington, J. D. 1965. Organic production, turnover and mineral cycling in woodlands. *Biol. Rev.* 40:295-336.

Ovington, J. D., and J. S. Olson. 1970. Biomass and chemical content of El Verde lower montane rain forest plants. In *A tropical rain forest,* ed. H. T. Odum and R. F. Pigeon, pp. H53-75. USAEC.

Pereira, H. 1967. Effects of land use on the water and energy budgets of tropical watersheds. In *Forest Hydrology,* ed. W. E. Sopper and H. W. Lull, pp. 435-450. Oxford: Pergamon Press.

Pereira, H. C., M. Dagg, and P. H. Hosegood. 1962. The water balance of both treated and control valleys. *East African Agri. and For. Jour. Special Issue* 27:36-41.

Pires, J. M., T. H. Dobzhansky, and G. A. Black. 1953. An estimate of the number of species of trees in an Amazonian forest community. *Bot. Gaz.* 114:467-477.

Pittier, H. 1918. Our present knowledge of the forest formations of the isthmus of Panama. *J. Forestry* 16:76-84.

Pomeroy, L. R. 1970. The strategy of mineral cycling. *Ann. Rev. Ecology and Systematics* 1:171-190.

Popenoe, H. 1959. The influence of the shifting cultivation cycle on soil properties in Central America. *Proc. Ninth Pacific Sci. Cong.* 7:72-77.

Popenoe, H. 1966. Soil management in land settlement and development of the tropics. Mimeo. presented at a Seminar on "Problems of land settlement and development in the tropics." Univ. of Michigan.

Portig. W. H. 1965. Central American rainfall. *Geographical Review* 55:69-90.

Rajan. S. V. G. and R. S. Murthy. 1971. Trends in rock weathering in the southern part of peninsula India—its expression in morphogenesis of soils. In *Soils and tropical weathering; Proceedings of the 1969 Bandung Symposium,* pp. 66-72. Paris: UNESCO.

Rankama, K., and T. G. Sahama. 1959. *Geochemistry.* Chicago: Univ. of Chicago Press.

Reichle, D. E. 1968. Relation of body size to food intake, oxygen consumption, and trace element metabolism in forest floor arthropods. *Ecology* 49:538-542.

Reichle, D. E., and D. A. Crossley, Jr. 1967. Investigation of heterotrophic productivity in forest insect communities. In *Secondary Productivity of Terrestrial Ecosystems,* ed. K. Petrusewicz, pp. 563-587. Warsaw.

Richards, P. W. 1939. Ecological studies on the rain forest of southern Nigeria. I. The structure and floristic compositions of the primary forest. *J. Ecology* 27:1-53.

Richards, P. W. 1945. The floristic composition of primary tropical rain forest. *Biol. Rev.* 20:1-13.

Richards, P. W. 1952. *The tropical rain forest: an ecological study.* Cambridge: University Press.

Robertson, R. A., and G. E. Davies. 1965. Quantities of plant nutrients in heather ecosystems. *J. Appl. Ecol.* 2:211-219.

Rodin, L. E., and N. I. Bazilevich. 1967. *Production and mineral cycling in terrestrial vegetation.* Edinburgh: Oliver and Boyd.

Rozanov, B. G., and I. M. Rozanova. 1964. Biological cycle of bamboo (Bambusa spp.) nutrients in the tropical forests of Burma. *Bot. Zhur.* (Moscow) 49:348-357.

Rosevear, D. R. 1947. Mangrove swamps. *Farm and Forest* 3:23-30.

Ross, R. 1954. Ecological studies on the rain forest of southern Nigeria. III. Secondary succession in the Shasha forest reserve. *J. Ecol.* 42:259-282.

Sabhasri, S., C. Khemnark, S. Aksornkoae, and P. Ratisoonthorn. 1968. Primary production in dry-evergreen forest at Sakaerat Amphoe Pak Thong Chai, Changwat Nakhon Ratchasima I. Estimation of biomass and distribution amongst various organs. Report 1, Ecosystem study of tropical dry-evergreen forest. Bangkok. Mimeo.

Salt, G. 1952. The arthropod population of the soil in some east African pastures. *Bull. Entomol. Res.* 43:203-220.

Sauer, C. O. 1966. *The early Spanish main.* Berkeley and Los Angeles: Univ. of Calif. Press.

Semago, W. T., and A. J. Nash. 1962. Interception of precipitation by a hardwood forest floor in the Missouri Ozarks. Univ. of Missouri. Agr. Exp. Sta. Res. Bull. no. 796.

Seth, S. K., O. N. Kaul, and A. C. Gupta. 1963. Some observations on nutrition cycle and return of nutrients in plantations at New forest. *Ind. For.* 89:90-98.

Sexton, O. J., H. F. Heatwole, and E. H. Meseth. 1963. Seasonal population changes in the lizard, *Anolis. limifrons,* in Panama. *Amer. Midl. Nat.* 69(2):482-491.

Shanks, R. E., E. E. C. Clebsch, and H. R. DeSelm. 1961. Estimates of standing crop and cycling rate of minerals in Applachian ecosystems. AIBS Meeting, Purdue Univ. Aug. 30, 1961. Mimeo.

Sherman, G. D. 1971. Mineral weathering in relation to utilization of soils. In *Soils and tropical weathering. Proceedings of the 1969 Bandung Symposium,* pp. 51-56. Paris: UNESCO.

Singh, K. P. 1967. Production, chemical composition and decomposition of leaf litter, and soil properties in the deciduous forest communities at Varanasi. Ph.D. dissertation, Banaras Hindu Univ.

Singh, K. P. 1968. Litter production and nutrient turnover in deciduous forest of Varanasi. *Proc. Symp. Recent Adv. Trop. Ecol.* pp. 655-665.

Snedaker, S. and J. Gamble. 1969. Compositional analysis of selected second growth species from lowland Guatemala and Panama. In *Symposium on sea-level canal bioenvironmental studies.* IOCS memorandum BMI-24, report 9.

Snow, D. W. 1965. A possible selective factor in the evolution of fruiting seasons in tropical forest. *Oikos* 15(11):274-281.

Stark, N. 1970. The nutrient content of plants and soils from Brazil and Surinam. *Biotropica* 2(1):51-60.

Stark, N. 1971a. Nutrient cycling I. Nutrient distribution in some Amazonian soils. *Tropical Ecology* 12:24-50.

Stark, N. 1971b. Nutrient cycling II. Nutrient distribution in Amazonian vegetation. *Tropical Ecology* 12:177-201.

Strickland, A. H. 1947. The soil fauna of two contrasted plots of land in Trinidad, British West Indies. *Jour. Anim. Ecol.* 16(1):1-10.

Tadaki, Y. 1966. Some discussions on the leaf biomass of forest stands and trees. *Bull. Gov. For. Exp. Sta.* (Tokyo) 184: 135-161.

Templeton, W. L., J. M. Dean, D. G. Watson and L. A. Rancitelli. 1969. Freshwater ecological studies in Panama and Columbia. In *Symposium on sea level canal bioenvironmental studies.* IOCS memorandum BMI-24, report 16.

Tergas, L. E. 1965. Correlation of nutrient availability in soil and uptake by native vegetation in the humid tropics. M.S. thesis. Univ. of Florida, Gainesville.

Tropical Test Center. 1966. Environment data base for regional studies in the humid tropics. Semiannual report No. 1 and 2. USATECOM project no. 9-4-9913-01.

Tsutsumi, T., M. Kan, and C. Khemanark. 1966. The amount of plant nutrients and their circulation in the forest soils in Thailand—carbon and nitrogen contents and some physical properties of the forest soils. (In Japanese.) *S. E. Asian Studies* 4:327-366.

Tsutsumi, T., M. Kan, and C. Khemanark. 1967. The amount of plant nutrients and their circulation in the forest soils in Thailand: the amount of bases, phosphorus and their circulation. (In Japanese.) *S. E. Asian Studies* 4:897-928.

Tsutsumi, T., T. Kawahara, and T. Shidei. 1968. The circulation of nutrients in forest ecosystem. (1) On the amount of nutrients contained in the above-ground parts of single tree and of stand. *J. Jap. For. Soc.* 50(3):66-74.

Tukey, H. B., Jr., and H. B. Tukey, Sr. 1962. The loss of organic and inorganic materials by leaching from leaves and other above-ground plant parts. In *Radio-isotopes in soil and plant nutrition studies,* pp. 289-302. Vienna.

Underwood, E. J. 1956. *Trace elements in human and animal nutrition.* New York: Academic Press.

Walter, H., and H. Lieth 1960-67. Klimadiagram-weltatlas. Jena: V. B. Gustav Fisher Verlag.

West, R. C. 1956. Mangrove swamps of the Pacific coast of Columbia. *Ann. Assoc. Amer. Geogr.* 46(1):98-121.

Wiegert, R. G. 1970. Effects of ionizing radiation on leaf fall decomposition, and the litter microarthropods of a Montane rain forest. In *A tropical rain forest,* ed. H. T. Odum and R. F. Pigeon, pp. H89-100. USAEC.

Wiegert, R. G., and P. Murphy. 1970. Effect of season, species, and location on the disappearance rate of leaf litter in a Puerto Rican rain forest. In *A tropical rain forest,* ed. H. T. Odum and R. F. Pigeon, pp. H101-104. USAEC.

Williams, E. C. 1941. An ecological study of the floor fauna of the Panama rain forest. *Bull. Chicago Acad. Sci.* 6:63-124.

Witkamp, M., and J. S. Olson. 1963. Breakdown of confined and nonconfined oak litter. Mimeo. Oak Ridge, Tenn.

Woods, F. W., and C. M. Gallegos. 1970. Litter accumulation in selected forests of the Republic of Panama. *Biotropica* 2(1):46-50.

Woodwell, G. M., and R. H. Whittaker. 1968. Primary production and the cation budget of the Brookhaven forest. In *Symposium on primary productivity and mineral cycling in natural ecosystems,* pp. 151-166. Orano: Univ. of Maine Press.

Zeuthen, E. 1953. Oxygen uptake as related to body size in organisms. *Quart. Rev. Biol.* 28(1):1-11.

# Index